INSTABILITIES in LASER-MATTER INTERACTION

Sergei I. Anisimov
Viktor A. Khokhlov
L.D. Landau Institute for Theoretical Physics
Russian Academy for Science
Moscow

CRC Press
Boca Raton   London   Tokyo

Library of Congress Cataloging-in-Publication Data

Anisimov, S. I.
  Instabilities in Laser-matter interaction / Sergei I. Anisimov, Viktor A. Khokhlov.
    p. cm.
  Includes bibliographical references and index.
  ISBN 0-8493-8660-8
  1. Laser beams. 2. Matter—Effect of radiation on. I. Khokhlov, Viktor A. II. Title.
QC688.A64 1995
620.1′1228—dc20
DNLM/DLC
for Library of Congress   94-23471
                                                                                          CIP

This book contains information obtained from authentic and highly regarded sources. Reprinted material is quoted with permission, and sources are indicated. A wide variety of references are listed. Reasonable efforts have been made to publish reliable data and information, but the author and the publisher cannot assume responsibility for the validity of all materials or for the consequences of their use.

Neither this book nor any part may be reproduced or transmitted in any form or by any means, electronic or mechanical, including photocopying, microfilming, and recording, or by any information storage or retrieval system, without prior permission in writing from the publisher.

CRC Press, Inc.'s consent does not extend to copying for general distribution, for promotion, for creating new works, or for resale. Specific permission must be obtained in writing from CRC Press for such copying.

Direct all inquiries to CRC Press, Inc., 2000 Corporate Blvd., N.W., Boca Raton, Florida 33431.

© 1995 by CRC Press, Inc.

No claim to original U.S. Government works
International Standard Book Number 0-8493-8660-8
Library of Congress Card Number 94-23471
Printed in the United States of America   2 3 4 5 6 7 8 9 0
Printed on acid-free paper

Contents

PREFACE   5

CHAPTER 1. INTERACTIONS OF LASER RADIATION
WITH HIGHLY ABSORBING MATERIALS   9
1.1 Laser heating of highly absorbing solids without phase change .   9
1.2 Laser-induced melting   13
1.3 Laser-induced evaporation   15
1.4 Numerical simulation of laser-produced melting and evaporation   18

CHAPTER 2. INTERACTION OF LASER RADIATION
WITH DIELECTRICS   25
2.1 Optical breakdown of transparent dielectrics   25
2.2 Stationary evaporation waves in transparent dielectrics   32
2.3 UV-laser ablation of polymers   35

CHAPTER 3. PROCESSES IN THE VAPOR PLUME   45
3.1 Hydrodynamic boundary conditions for strong evaporation . .   45
3.2 Condensation in expanding vapor   48
3.3 Dynamics of vapor expansion: one-dimensional flow   49
3.4 Dynamics of vapor expansion: spatial structure of the vapor plume   51
3.5 Instability of the vapor-plume expansion into a vacuum   54
3.6 Dynamics of vapor expansion into an ambient gas   56

CHAPTER 4. EFFECTS OF ULTRASHORT LASER PULSES   63
4.1 Electron and lattice temperature dynamics in a metal heated by ultrashort laser pulses   63
4.2 Anomalous electron and light emission from metals heated by ultrashort laser pulses   65

CHAPTER 5. INSTABILITY OF SUBLIMATION WAVES
IN SOLIDS: LINEAR THEORY   73
5.1 Corrugation instability of stationary evaporation waves in highly absorbing materials   73
5.2 Corrugation instability of nonstationary evaporation waves in highly absorbing materials   79

5.3   Instability of plane stationary evaporation waves in dielectrics . 82
5.4   Stability analysis of plane stationary ablation waves
      in polymers 88

CHAPTER 6. NONLINEAR EVOLUTION OF INSTABILITY IN LASER EVAPORATION 93
6.1   Slightly supercritical structures in laser sublimation waves . . . 93
6.2   Strongly supercritical regimes of laser evaporation.
      "Spin" sublimation of solids. 99
6.3   Nonlinear stages of instability of evaporation waves
      in dielectrics 101

CHAPTER 7. INSTABILITIES IN THE LASER INTERACTION WITH LIQUIDS 109
7.1   Thermocapillary instability of a liquid heated by laser
      radiation 109
7.2   Corrugation instability of plane evaporation waves in liquids . . 114

CHAPTER 8. INSTABILITIES CONNECTED WITH LASER-INDUCED CHEMICAL REACTIONS 125
8.1   Instability of laser-stimulated surface oxidation of metals.
      Spatially uniform temperature fields 125
8.2   Instability of surface oxidation of metals:
      Two-dimensional effects 128
8.3   Laser-initiated surface combustion waves 131

CONCLUSION 135

INDEX 143

PREFACE

The first experimental studies of laser–matter interaction were conducted in the early 1960s when the first high-power lasers were created. In these works, mainly thermal effects were studied including heating, melting, evaporation of condensed matter, and thermionic electron emission. When higher laser intensities were reached, it became possible to study the optical breakdown of gas and solid dielectrics. In the 1960s, the first theoretical models of laser–matter interaction also were proposed. The simplest models were related to moderate laser intensity ranges ($I \simeq 10^5$–10^9 W/cm^2) and took into account the energy transfer in condensed matter, the kinetics of phase transitions, and the dynamics of evaporated material expansion.[1-4] A survey of the results obtained during the initial period of laser–matter interaction studies can be found in Ready[5] and Anisimov et al.[6] Despite many simplifications, theoretical models[1-4] provide a basic outline of the actual interaction of laser radiation with solids at moderate laser intensities. However, it appeared to be difficult to reach a quantitative agreement between theory and experiments. In particular, the detailed analysis of phase composition and energy balance in laser ablation of metals[1,5,7] shows that a considerable part of the ablated material is in the form of liquid droplets, and the specific ablation energy—equal to the ratio of the incident laser energy to the mass of the material removed from the target—is several times less than the specific heat of vaporization. These facts cannot be explained by simple theoretical models. Another manifestation of the very complicated character of actual laser–matter interaction processes and the inadequacies of simple theoretical models follows from studies of target reflectivity changes during laser irradiation.[5,8] The experiments show that total reflectivity and particularly its specular part, is reduced considerably during a laser pulse. This reduction cannot be explained easily by a decrease in surface layer conductivity due to heating. Finally, such important characteristics as the phase composition of the ablated material, the specific ablation energy, and the recoil momentum transferred to the target are extremely sensitive to the specifics of the laser pulse structure and focusing conditions.[9]

These facts can be understood from a single viewpoint if we suppose that laser-induced evaporation of condensed materials can be unstable under certain conditions. The first example of laser-produced evaporation instability, studied by Anisimov, Tribel'skii, and Epel'baum[10] was the corrugation instability of a plane stationary vaporization front in highly absorbing solids. This instability is driven by the temperature gradient in the vicinity of the phase-transition front. A similar instability mode appears in laser-induced evaporation of transparent dielectrics. Furthermore, the thermal and avalanche breakdown of transparent dielectrics also can be considered to be a particular mode of instability of these materials in a strong electromagnetic field. Many other examples of instability have been studied during recent years.[11-13] These studies clearly demonstrate that most of the laser-matter interaction processes are unstable.

In the present book, a review is given of thermal and hydrodynamic instabilities appearing in laser-matter interactions at moderate intensities. These instabilities also can arise in the processes of heating of condensed matter with electron and ion beams, shock waves, etc. Nonresonant interactions are the subject of this book; to include resonant electromagnetic modes related to the generation of surface electromagnetic waves would require a second book. Note also that these phenomena are not characteristic of the interaction of electron or ion beams with solids. A comprehensive review of the phenomena associated with surface electromagnetic waves can be found in references 11–13. We will not consider here problems of laser-driven inertial confinement fusion; instead, we will restrict our attention to the range of low and moderate laser intensities that are important for technological applications of lasers.

This book is constructed as follows: In the first four chapters, we give a survey of the basic processes of nonresonant laser-matter interactions. Laser-induced breakdown of transparent dielectrics is considered in Chapter 2, as an example of instability of "thermal explosion" type (see Frank-Kamenetskii[14]). Chapter 4 contains a brief analysis of the effects produced by ultrashort (picosecond and femtosecond) laser pulses. This problem recently has attracted extensive interest due to the important applications of ultrafast laser technology. The problems of stability of thermal processes induced by ultrashort laser pulses have not yet been studied.

In chapters 5 to 8 we consider thermal and hydrodynamic instabilities. This provides a theoretical background to the interpretation of experimental results and an understanding of the effect of instabilities on the processes of laser technology.

We would like to note that the instabilities discussed in this book are important for the interpretation of laser interaction experiments and technological applications of lasers. The instabilities are also of interest from the standpoint of fundamental physics. There are many examples of physical systems whose stable steady states have a symmetry that does not correspond to the symmetry of external conditions. In these cases, it is customary to discuss spontaneous symmetry breaking. Many examples of this general phenomenon are known in the fields of atomic and molecular physics, solid state physics, hydrodynamics, and quantum field theory.[15-17] Laser-matter interactions give additional examples of symmetry breaking due to instability. The analysis of the symmetry breaking and structure formation is a part of the general theory of self-organization, which has been given notice during the past 20 years. The concepts and methods of this theory can be employed, as we will see later, in laser–matter interaction studies.

This book is devoted mainly to the theoretical analysis of laser–matter interaction and instability phenomena. This analysis is based primarily on the results of studies carried out by Russian physicists during the past 10–15 years. These results were published primarily in Russian journals and were delayed in reaching western readers.

The authors are grateful to their friends and colleagues: D. Bäuerle, A. M. Bonch-Bruevich, M. N. Libenson, B. S. Luk'yanchuk, J. Meyer-ter-Vehn, M. I. Tribel'skii, and S. Witkowski for helpful discussions and criticism during the preparation of this book. One of the authors (S. A.) acknowledges with gratitude the kind hospitality of the Max-Planck-Institute for Quantum Optics (Garching, Germany) and the support of the Alexander von Humboldt Foundation.

REFERENCES

1. Anisimov, S. I., Bonch-Bruevich, A. M., Elyashevich, M. A., Imas, Ya. A., Pavlenko, N. A., and Romanov, G. S., Effect of powerful light fluxes on metals, *Sov. Phys.–Tech. Phys.*, **11**, 945, 1967

2. Ready J. F., Effects due to absorption of laser radiation, *Journ. Appl. Phys.*, **36**, 462, 1965

3. Anisimov, S. I., Evaporation of light-absorbing metals, *High Temperature*, **6**, 110, 1968

4. Aleshin, I. V., Bonch-Bruevich, A. M., Imas, Ya. A., Libenson M. N., Rubanova G. M., and Salyadinov, V. S., Laser-induced evaporation of nonlinearly absorbing dielectrics, *Sov. Phys.–Tech. Phys.*, **22**, 1400, 1977

5. Ready, J. F., *Effects of High-Power Laser Radiation*, Academic Press, New York, 1971

6. Anisimov, S. I., Imas Ya. A., Romanov, G. S., and Khodyko, Yu. V., *Action of High-Power Radiation on Metals*, National Technical Information Service, Springfield, Virginia, 1971

7. Panteleev, V. V. and Yankovskii, A. A., Use of laser beams for vaporizing materials in spectral analysis, *Journ. Appl. Spectroscopy*, **3**, 260, 1965

8. Bonch-Bruevich, A. M., Imas, Ya. A., Romanov, G. S., Libenson, M. N., and Mal'tsev, L. N., Effect of a laser pulse on the reflecting power of metals, *Sov. Phys.–Tech. Phys.*, **13**, 640, 1968

9. Bass, M., Nassar, M. A., and Swimm, R. T., Impulse coupling to aluminum resulting from Nd: glass laser irradiation induced material removal, *Journ. Appl. Phys.*, **61**, 1137, 1987

10. Anisimov, S. I., Tribelskii, M. I., and Epelbaum, Ya. G., Instability of a plane evaporation front in the interaction of laser radiation with a medium, *Sov. Phys.-JETP*, **51**, 802, 1980

11. Bonch-Bruevich, A. M. and Libenson, M. N., Laser-induced surface polaritons and optical breakdown, In *Nonlinear Surface Electromagnetic Phenomena*, Ed. by Ponath, H.-E. and Stegman, G. I., Elsevier Science Publ., 1991, 561

12. Bonch-Bruevich, A. M., Libenson, M. N., Makin, V. S., and Trubaev, V. V., Surface electromagnetic waves in optics, *Optical Engineering*, **31**, 718, 1992

13. Akhmanov, S. A., Emel'yanov, V. I., Koroteev, N. I., and Seminogov, V. N., Interaction of powerful laser radiation with the surfaces of semiconductors and metals: nonlinear optical effects and nonlinear optical diagnostics, *Sov. Phys.-Uspekhi*, **28**, 1084, 1985

14. Frank-Kamenetskii, D. A., *Diffusion and Heat Transfer in Chemical Kinetics*, Plenum Press, New York, 1969

15. Elliott, J. P. and Dawber, P. G., *Symmetry in Physics*, **12**, Macmillan Press, London, 1979

16. Haken, H., *Synergetics*, Springer-Verlag, New York, 1978

17. Anisimov, S. I. and Tribel'skii, M. I., Instability and spontaneous symmetry breaking in macroscopic laser–matter interaction, *Sov. Sci. Rev., A Phys.*, **8**, 259, 1987

Chapter 1

INTERACTIONS OF LASER RADIATION WITH HIGHLY ABSORBING MATERIALS

Generally, the entire field of laser-matter interactions can be divided into two parts: (1) resonant and (2) nonresonant interactions. The first part includes selective excitation processes in systems with discrete spectra, multiphoton ionization of atoms and molecules, multiphoton photoelectron emission, and nonlinear optical phenomena in solids and liquids. Such factors as coherence, polarization, and spectral characteristics of laser radiation affect these excitation processes. Laser-induced thermochemical and thermophysical phenomena can be related to nonresonant interactions, as well as laser breakdown of gases and solids and laser plasma generation. For this group of phenomena, the coherence and spectral characteristics of the laser radiation are not very important. Principal effects are connected, in this case, with the heating of a material, and the important factors are the laser intensity, pulse duration, and focusing conditions. In this book, we shall discuss mainly the thermal effects of laser radiation. The emphasis will be on laser heating, melting, and vaporization of solid materials and the related instability of these processes. We consider also the dynamics of vapor plume expansion, with an emphasis on hydrodynamic instabilities of vapor expansion.

1.1 Laser heating of highly absorbing solids without phase change

When laser radiation intensity is rather low, no phase transition occurs, and the only effect of laser absorption is heating of the material. We first consider the laser heating of metals, the most important materials with high optical absorption. In metals, the laser radiation is absorbed by "free" electrons, and energy transfer in metals is also due to electron heat conduction. The temperature field is described by the standard heat conduction equation:

$$\rho c \partial T/\partial t = \text{div}\,(\kappa \nabla T) + Q(\mathbf{r}, t) \tag{1.1}$$

where T is the temperature, κ is the thermal conductivity, c is the specific heat, ρ is the density, and $Q(\mathbf{r}, t)$ is the heat production (per unit volume per unit time) due to the laser radiation absorption. In the general case, c and κ are temperature-dependent, and Equation (1.1) is nonlinear. For metals, however, at temperatures above the Debye temperature, c and κ can be taken as constants.[1,2]

It should be noted that the macroscopic approach to the problem of laser heating is justified when the characteristic time and space scales of the temperature field are much larger than the mean free time and mean free path of electrons, respectively. These conditions are fulfilled in the case of nanosecond (and longer) laser pulses. However, in the case of picosecond and subpicosecond pulses, the above approach should be modified.

The heat source $Q(\mathbf{r}, t)$ in Equation (1.1) depends on the laser pulse parameters and on the optical properties of the material irradiated. In many practical cases, the Drude-Zener theory[2,3] provides an adequate description of the optical properties of materials whose optical absorption is mainly due to free electrons. Solid and liquid metals are typical examples of such materials. The Drude-Zener theory leads to the following expression for $Q(\mathbf{r}, t)$:

$$Q = A\mu I(x, y, t)\exp(-\mu z) \tag{1.2}$$

where A is the surface absorptivity, μ is the absorption coefficient of the material, and $I(x, y, t)$ is laser radiation intensity at the material surface ($z = 0$). The absorption coefficient, μ, is related to the skin depth, δ, by $\mu = 2/\delta$. According to Sokolov[2] and Libenson et al.[4] μ is independent of temperature, while $A = 1 - R$ (R is the surface reflectivity) is a linear function of the surface temperature:

$$A(T) = A_0 + A_1(T - T_0) \tag{1.3}$$

where A_0 is the surface absorptivity at room temperature, T_0. This temperature dependence of A results from the fact that A is proportional to the electron-phonon collision frequency which, in turn, is proportional to the crystal lattice temperature.

Boundary conditions for Equation (1.1) can be written in the following form. At a large distance from the material surface, as $z \to \infty$

$$T \to T_0 = \text{constant} \tag{1.4}$$

(in the majority of cases, T_0 can be set equal to zero). On the material surface, $z = 0$, the continuity condition for the energy flux should be imposed. It reads as:

$$-\kappa \partial T/\partial z = q_c + q_r$$

where q_c and q_r are energy fluxes due to convection and radiation, respectively. In the case under consideration (no phase transition) the temperature of the

CHAPTER 1. INTERACTIONS OF LASER RADIATION

surface is not very high, and the energy losses are negligible, compared to the laser energy flux. An estimate of the region of laser-pulse parameters in which the energy losses are negligible is given by Ready.[1] We can thus set:

$$\partial T/\partial z = 0 \quad \text{at } z = 0 \tag{1.5}$$

The initial condition for Equation (1.1) reads:

$$T(\mathbf{r}, 0) = T_0 \tag{1.6}$$

In many cases of practical interest, the transverse dimensions of the laser focusing spot are large compared to the thickness of the heated layer, and the heat conduction problem (Equations (1.1) and (1.4) through (1.6)) can be considered one-dimensional.

Note that the heat conduction problem (1.1) with $Q(\mathbf{r}, t)$ defined by Equations (1.2) and (1.3) is linear and can be solved using standard methods.[5] In the simplest case of time- and coordinate-independent laser intensity, $I(x, y, t) = I_0 = $ constant, this problem has been solved by Libenson et al.[4] To obtain the solution, we apply the Laplace transform, defined as:

$$u(z, p) = \int_0^\infty T(z, t) \exp(-pt)\, dt$$

to Equation (1.1) and to the boundary conditions (1.4) and (1.5). The ordinary differential equation that results from the Laplace transformation can be solved readily to obtain:

$$\begin{aligned} u(z,p) &= (\mu I_0/c\rho)\left[(A_1 u_0 p + A_0)/p(p - \chi\mu^2)\right] \\ &\quad \times \left[\exp(-\mu z) - (\mu/\beta)\exp(-\beta z)\right] \end{aligned} \tag{1.7}$$

Here, $u_0 = u(0, p)$, $\beta = (p/\chi)^{1/2}$, and $\chi = \kappa/c\rho$ is the heat diffusivity. The temperature, $T(z, t)$, is calculated by applying the inverse Laplace transform to Equation (1.7). To avoid cumbersome formulas, we perform the calculation only for the surface temperature, $T(0, t)$. Solving Equation (1.7) for u_0, we obtain

$$u_0 = [I_0 A_0 \mu/\rho c p(\alpha_1 - \alpha_2)]\left[\frac{1}{\sqrt{p} - \alpha_1} - \frac{1}{\sqrt{p} - \alpha_2}\right] \tag{1.8}$$

Here $\alpha_{1,2} = (\mu\sqrt{\chi}/2)(-1 \pm \sqrt{1+S})$, and $S = 4I_0 A_1/\mu\kappa$. The parameter S is proportional to the laser intensity, and in the case under consideration, it is relatively small. For typical metals ($A_1 \simeq 10^{-4}$ K^{-1}, $\mu \simeq 10^5$ cm^{-1}, $\kappa = 3$ W/cm K), $S \simeq 10^{-2}$ at laser intensity 10^7 W/cm^2. Assuming $S \ll 1$ and performing the inverse Laplace transform, we obtain from (1.8):

$$\begin{aligned} T(0, t) &= T_0 + (I_0 A_0/\kappa\mu)\left[\exp(\mu^2 \chi t)\operatorname{erfc}(\mu\sqrt{\chi t}) - 1\right] \\ &\quad + (A_0/A_1)\left[\exp(\gamma t)\operatorname{erfc}(-\sqrt{\gamma t}) - 1\right] \end{aligned} \tag{1.9}$$

where

$$\operatorname{erfc}(x) = (2/\sqrt{\pi}) \int_x^\infty \exp(-y^2)\, dy, \quad \text{and} \quad \gamma = I_0^2 A_1^2/\kappa c\rho$$

In the majority of calculations regarding the laser heating of solid surfaces, the temperature dependence of the surface reflectivity usually is neglected. This results in the following equation:

$$T(0,t) = T_0 + (I_0 A_0/\kappa\mu)\left[2\mu\sqrt{\chi t}/\sqrt{\pi} + \exp(\mu^2\chi t)\,\text{erfc}(\mu\sqrt{\chi t}) - 1\right] \quad (1.10)$$

which follows from (1.9) if we set $A_1 = 0$. It is easy to see that Equations (1.9) and (1.10) give almost the same values of the surface temperature if $\gamma t \ll 1$. However, when $t \gg \gamma^{-1}$, the changes in surface reflectivity become important. This leads to exponential time growth of the surface temperature at large t:

$$T(0,t) \simeq T_0 + 2(A_0/A_1)\exp(\gamma t) \quad (1.11)$$

Note that when $S \gg 1$, only the regime with exponential temperature growth exists, described by the following equation:

$$T(0,t) = (A_0/A_1)\left[\exp(I_0 A_1 \mu t/c) - 1\right]$$

In many cases of practical interest, the skin depth is much smaller than the spatial scale of the temperature field, and we can consider light absorption as a surface process. Thus, we can set $Q = 0$ in (1.1) and change the boundary condition (1.5) at $z = 0$ to:

$$-\kappa \partial T/\partial z = A(T)I_0 \quad (1.12)$$

The solution of the last problem can be obtained formally from (1.9) by setting $\mu \to \infty$. The result is given by the equation:

$$T(0,t) = (A_0/A_1)\left[\exp(\gamma t)\,\text{erfc}(-\sqrt{\gamma t}) - 1\right]$$

which coincides with (1.11) in the limiting case $\gamma t \gg 1$.

The spatial temperature profile can be calculated by taking the inverse Laplace transform of (1.7). Since the result for a general case is cumbersome, we will consider only the simplest case of constant reflectivity, $A_1 = 0$. The temperature field is given by:[5]

$$\begin{aligned}T(z,t) &= (2I_0 A_0/\kappa)\sqrt{\chi t}\,\text{ierfc}(z/2\sqrt{\chi t}) - (I_0 A_0/\mu\kappa)\exp(-\mu z) \\ &+ (I_0 A_0/2\mu\kappa)\exp(\mu^2\chi t - \mu z)\,\text{erfc}(\mu\sqrt{\chi t} - z/2\sqrt{\chi t}) \\ &+ (I_0 A_0/2\mu\kappa)\exp(\mu^2\chi t + \mu z)\,\text{erfc}(\mu\sqrt{\chi t} + z/2\sqrt{\chi t})\end{aligned} \quad (1.13)$$

where

$$\text{ierfc}(x) = \int_x^\infty \text{erfc}(y)\,dy. \quad \text{For } z = 0,\ (1.13)\text{ reduces to }(1.10).$$

As μ tends to infinity (surface absorption), we have from (1.13) the following simple result:

$$T(z,t) = (2I_0 A_0/\kappa)\sqrt{\chi t}\,\text{ierfc}(z/2\sqrt{\chi t})$$

We have considered the one-dimensional temperature field produced by laser heating of a highly absorbing solid. Note that this field is nonstationary:

the surface temperature increases over a long time as $\exp(\gamma t)$ (at $S \ll 1$), or as $\exp(\sqrt{\chi\gamma}\mu t)$ (at $S \gg 1$), and the thickness of the heated surface layer increases as $\sqrt{\chi t}$.

A one-dimensional approximation is valid when the heated layer thickness $h \propto \sqrt{\chi t}$ is much smaller than the laser spot size R. It is clear that this condition is broken when $t \geq R^2/\chi$. Thus, when the laser pulse length is larger than R^2/χ, the temperature distribution becomes two- or three-dimensional. This situation occurs, for example, when a material is heated by a continuous laser. In this case, the spatial scale of temperature distribution is much larger than the skin depth, and the light absorption may be considered as a surface process. We can, therefore, set $Q = 0$ in Equation (1.1) and use the boundary condition similar to (1.12) with laser flux intensity I_0, localized within the laser focusing spot. It should be noted that energy loss due to convection and radiation can be no longer neglected, since for long laser pulses, a large area around the laser spot becomes hot and contributes to heat loss. In this case, the heat loss from the surface $z = 0$ may reach the same order of magnitude as the laser power absorbed. The linear heat conduction equation (1.1) with $Q = 0$ and linear boundary condition at $z = 0$, can be solved easily using standard methods.[5] Examples of multidimensional temperature field calculations can be found in Ready[1] and Carslaw and Jaeger.[5] We will not consider the details of these calculations; note that for a time-independent laser intensity a steady-state temperature profile can be reached in the multidimensional case.

The above analysis of heat transfer in a laser-heated solid contains a number of simplifications. We did not consider the real temporal shape of a laser pulse, which can be important for quantitative calculation of the temperature. In the framework of linear heat conduction problems, the consideration of arbitrary laser pulse shape does not lead to any complications, but gives rise to more laborious calculations, which sometimes require the application of numerical methods. More serious difficulties may be related to the nonlinearity of the heat transfer problem that results from temperature dependence of thermophysical and optical characteristics of the material. Nonlinearity also can appear in the boundary condition if radiative energy loss is taken into account. Nonlinear heat conduction problems may be solved, generally, by approximate or numerical methods. Examples of such calculations can be found in Prokhorov et al.[6]

1.2 Laser-induced melting

We now turn to higher laser energy absorption and consider the melting of solids absorbing laser radiation. In practice, laser-induced melting usually is accompanied by vaporization. Since the vaporization rate depends strongly on the temperature, melting without substantial vaporization can be observed only in a very narrow range of laser parameters. As noted by A. I. Shal'nikov, this range depends on the saturated vapor pressure of a particular material at its melting point.[7] It is important to note that for different materials, this parameter varies by many orders of magnitude. For example, the saturated vapor pressure at the melting point equals 9.8×10^{-3} bar for Cr, 5.7×10^{-2} bar for Br, and 2.4×10^{-1} bar for Gd; at the same time, for Ga this pressure is 9.2×10^{-41} bar, for Sn 5.7×10^{-26} bar and for S 2.6×10^{-25} bar.[8] It is clear that for the materials with rather high saturated vapor pressures at the melting point, pure melting without vaporization rarely is observed.

The simplest estimate of the mass of fused material can be obtained if one supposes that the melting front coincides with the melting temperature isotherm. In this estimate, the latent heat of fusion is neglected, which is not small, in most cases, in comparison with the total enthalpy of a material at the melting point. A more accurate calculation of melting front position can be made in terms of the so-called "problem of Stefan."[5] With this approach, the temperature fields in the solid and liquid phases are described by the heat conduction equations, and the boundary condition on the moving phase boundary is assumed to have the following form:

$$-\kappa_l \partial T_l/\partial n + \kappa_s \partial T_s/\partial n = V_n \rho L_m \qquad (1.14)$$

where $\partial/\partial n$ is the normal derivative, L_m is the latent heat of fusion, and V_n is the melting front velocity. The melting front is defined, in the framework of the "problem of Stefan," as an isothermal surface where the temperature is equal to the melting temperature, T_m. Note that the law of motion of the melting front is not known *a priori* and should be found as an eigenvalue in the course of solution of the heat transfer problem. An important example of such a calculation is given by Carslaw and Jaeger.[5] This is the well known Neumann's solution for the semi-infinite region, initially at a constant temperature $T_0 > T_m$. Physically, this solution corresponds to instantaneous heating of the solid material and may be considered, in principle, as a model for pulsed-laser produced melting. It should be noted, however, that Neumann's solution leads to a physically meaningless conclusion that the melting front velocity becomes infinite, $V_n = \infty$, at finite superheating, $(T_0 - T_m)c/L_m \geq 1$. This qualitatively incorrect result is the consequence of the basic assumption of the "problem of Stefan" that the melting front dynamics are governed by the energy flux only and are independent of the phase transition kinetics. It is clear that this simple assumption is applicable only if the melting front velocity is very small.

Melting kinetics at low superheating have been considered by Motorin and Musher.[9] Assuming the melting front velocity to be proportional to the difference of chemical potentials of solid and liquid phases, $\mu_s - \mu_l$, Motorin and Musher[9] obtained the following formula for V_n:

$$V_n = V_m \left[c(T - T_m)/L_m + (cT_m \sigma/\rho L_m^2)(1/R_1 + 1/R_2)\right] \qquad (1.15)$$

where σ is the surface energy density (surface tension) at the solid–liquid interface, R_1 and R_2 are the principal radii of curvature of the phase boundary, and V_m is an empirical constant that can be determined experimentally. In order of magnitude, V_m is close to the speed of sound. For copper, e.g., V_m equals 7×10^4 cm/s.[9] When phase transition kinetics are taken into account, the temperature at the melting front can no longer be assumed to be equal

CHAPTER 1. INTERACTIONS OF LASER RADIATION

to the normal melting point. The solid phase at the front has a temperature higher than that of the melting point, and the front velocity, as seen from (1.15), is proportional to the superheating. To calculate the melting front velocity and temperature, one should solve the heat conduction equations for the solid and liquid phases with boundary conditions (1.14) and (1.15) imposed on the front. These calculations (for a more general case with melting and evaporation kinetics both taken into account) were performed by Anisimov and Barsukov.[10] We shall discuss these calculations in Section 1.4.

It is important to note that the melting can be considered to be a surface process only when the superheating is sufficiently low. In the general case, the bulk formation of liquid phase nuclei (homogeneous nucleation) must be taken into account. There is a fundamental distinction between bulk and surface melting.[11] Usually, a liquid completely wets the surface of a solid phase of the same substance. Therefore, the formation of a liquid nucleus on the surface does not require work to be done to form a new surface. In other words, there is no energy barrier for the surface melting. On the contrary, the formation of a liquid nucleus within a crystal involves elastic deformations; this makes the bulk nucleation thermodynamically disadvantageous. This is true, of course, for any first-order phase transition in solids. The existence of the finite energy barrier for the bulk nucleation is used sometimes to explain the fact that first-order phase transitions do not begin on the phase equilibrium line.

It should be noted that the work required to form a nucleus of a new phase depends on the nucleus shape. For example, in the case of melting, a spherical nucleus formation can be expected to occur at the superheating of the order:[12]

$$\Delta = \frac{2cT_m KG\beta^2}{(4G + 3K)\rho_s L_m^2}$$

where K is the compressibility of the liquid phase; G is the shear modulus of the solid phase; and $\beta = (\rho_s - \rho_l)/\rho_s$, ρ_s, ρ_l is the density of solid and liquid phase, respectively. It was shown by Motorin and Musher,[12,13] however, that for disk-shaped nuclei, the effect of elastic deformations on the probability of nuclei formation is negligible. Thus, bulk melting can be expected to occur when the superheating is rather small. The value of superheating here is related to the finite growth time of the nucleus.

1.3 Laser-induced evaporation

As mentioned in the preceding section, melting without substantial vaporization can be produced only in a very narrow range of laser parameters. Usually, both phase transitions occur simultaneously under conditions typical for laser experiments. Note that the latent heat of vaporization is much larger than that of fusion (typically, by a factor of 20–50 times). Thus, evaporation plays the most important part in the energy balance.

In this book, we examine the interactions of laser radiations with matter at moderate laser intensities, $I \sim 10^5$–10^8 W/cm^2. At these intensities, the temperature is usually well below the thermodynamic critical point of the material. Under this condition, a sharp boundary exists between the condensed and gaseous phases, its thickness being on the order of interatomic distance, and the density of the vapor is several orders of magnitude less than in the condensed phase. In the majority of cases of practical interest, light absorption in the vapor is relatively small. To simplify our discussion we will begin with the assumption that the evaporation occurs in vacuum and the vapor does not absorb the laser radiation. Note that for some materials, the last assumption is not quite correct, especially near the upper boundary of the previous laser intensity range.

When the vapor plume does not absorb the laser light, the processes in the condensed and gaseous phases can be considered separately. As in the previous chapter, we shall describe the temperature field in the condensed phase by the heat conduction equation (1.1) with boundary conditions taking into account phase transition kinetics and energy balance. Note that the vaporization of condensed matter cannot be described, even in a very rough approximation, in terms of the "problem of Stefan" with a fixed phase transition temperature. Actually, a solid (liquid) evaporates at any temperature not equal to 0 K. The evaporation rate strongly depends on the temperature. This dependence, in the case under consideration, can be written in the form:[14,15]

$$V_n = V_0 \exp(-U/T) \tag{1.16}$$

where V_n is the normal component of the evaporation front velocity, $U = ML_v/k_B$, L_v is the latent heat of vaporization (per unit mass), k_B is Boltzmann's constant, M is the atomic mass, and V_0 is a constant whose value is of the order of the speed of sound in the condensed phase. We will show later how the evaporation rate is related to the saturated vapor pressure. Detailed data on saturated vapor pressures and evaporation rate of different materials can be found in references 8, 16, and 17.

Now we consider the simple problem of plane evaporation wave propagation.[14,18] Let the boundary between the condensed and gaseous phases be the plane and its law of motion due to condensed phase vaporization be $z = Z(t)$. Introducing new independent variables:

$$\zeta = \mu[z - Z(t)] \quad \text{and} \quad \tau = \mu^2 \chi t$$

we write the heat conduction equation (1.1) in the form

$$\partial T/\partial \tau = \partial^2 T/\partial \zeta^2 + \beta(\tau)\partial T/\partial \zeta + Q(\zeta,\tau) \tag{1.17}$$

where $\beta(\tau) = V/\mu\chi$; $V = dZ/dt$; $Q(\zeta,\tau) = K\exp(-\zeta)$; and $K = I(t)A/\mu\kappa$. The boundary conditions on the moving evaporation front $\zeta = 0$ are:

$$\partial T/\partial \zeta = (\Delta H/c)\beta(\tau) \quad \text{and} \quad \beta(\tau) = (V_0/\mu\kappa)\exp[-U/T(0,\tau)] \tag{1.18}$$

Here, ΔH is the difference in enthalpies of the gaseous and condensed phases, with respect to unit mass. It is approximately equal to $\Delta H \simeq L_v + k_B T/2M \simeq L_v$.

If the laser intensity I is constant, a steady-state solution of the boundary value problem (1.17) and (1.18) exists. It describes the stationary temperature

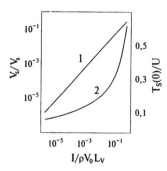

Figure 1.1: Temperature and velocity of the stationary evaporation fronts as functions of laser intensity.

profile $T_s(\zeta)$ propagating with constant velocity V_s through the solid (liquid) material. Assuming in (1.17) and (1.18) that $\beta(\tau) = \beta =$ constant; $T(\zeta, \tau) = T_s(\zeta)$; and $\partial T_s/\partial \tau \equiv 0$, we arrive at an ordinary differential equation, which can be solved as follows:

$$T_s(\zeta) = C_1 \exp(-\zeta) + C_2 \exp(-\beta\zeta) \tag{1.19}$$

where $C_1 = K/(\beta - 1)$ and $C_2 = -\Delta H/c - K/\beta(\beta - 1)$. It follows from (1.19) that the derivative, $\partial T_s/\partial \zeta$ at $\zeta = 0$, is positive and equal to $L_v V_s/c\kappa$. Since at $\zeta \to \infty$, $T_s(\zeta) \to 0$, and $\partial T_s/\partial \zeta < 0$, we conclude that the steady-state temperature at the phase boundary is lower than in the bulk of material. This fact plays a decisive role in the stability analysis, which we shall describe in Chapter 5.

If we set $\zeta = 0$ in (1.19) and use the equations for C_1, C_2, and K, we obtain the energy conservation equation, which can be written as:

$$I_{abs} = AI = V_s \rho [L_v + cT_s(0)] \tag{1.20}$$

where I_{abs} is the absorbed laser intensity. Taking the equation of vaporization kinetics (1.15) in the form:

$$V_s = V_0 \exp[-U/T_s(0)] \tag{1.21}$$

and solving the set of equations (1.20) and (1.21), we determine the stationary velocity, V_s, and the stationary temperature, $T_s(0)$, on the surface of the condensed phase. As an example, the results of such calculations in the case $c = 3k_B/M$ are presented in Figure 1.1. The dependence of the temperature and velocity of the stationary vaporization front on the laser intensity absorbed are given in references 1, 14, and 18 for different metals.

Let us consider now the transient process leading to the formation of a stationary evaporation wave. As Equation (1.19) shows, in a stationary wave, the spatial scale of temperature distribution near the solid-gas phase boundary is equal to $\max(\chi/V_s, \mu^{-1})$. To derive a very rough estimate of the time required to establish steady-state evaporation at constant absorbed laser intensity I_{abs}, we calculate the time of formation of the heated layer near the phase boundary. Since in the majority of cases $\chi/V_s > \mu^{-1}$, we see that this time is on the order of χ/V_s^2. The nature of the process of the establishment of steady motion of the phase boundary can be described as follows.

At the outset the phase boundary is at rest, and the entire laser energy flux absorbed is transferred by heat conduction into the bulk of material. Near the surface, a temperature gradient of the order I_{abs}/κ is created, which remains roughly constant, as long as the velocity of the boundary is low. Both the surface temperature and thickness of the heated layer near the surface increase with time as $\sqrt{t}$. Due to the very strong temperature dependence of the evaporation rate, the evaporation front velocity remains small, in comparison with V_s, until the temperature attains value very close to $T_s(0)$. Thus, throughout the period that lasts on the order of χ/V_s^2, the phase boundary remains practically at rest; for the remainder of the laser-pulse duration, the boundary moves with constant velocity V_s. It is clear that if the laser-pulse length is shorter than χ/V_s^2, then there is no appreciable evaporation of the material at all. Hence, we can estimate the critical laser intensity at which significant evaporation begins to be:

$$I_{abs}^* \simeq \rho L_m \sqrt{\chi/\tau_l}$$

where τ_l is the laser pulse duration. A similar qualitative analysis shows that in the case of $\chi/V_s \ll \mu^{-1}$, the time required to reach a steady-state evaporation regime is on the order of $(\mu V_s)^{-1}$. A more detailed analysis of the transient process leading to the formation of the stationary evaporation wave is presented in references 14 and 19–21. It is interesting to note that under certain conditions, oscillatory changes of the surface temperature and vaporization front velocity during the transient process may be observed.[21] This oscillatory behavior is the result of the volume nature of the laser radiation absorption, and it disappears if the absorption takes place at the surface.

We implicitly assumed that the surface reflectivity is constant. The effect of surface reflectivity change on the characteristics of the steady-state evaporation wave and its formation time was considered by Anisimov et al.[20] We shall not specify these calculations[20] because they were derived for the model dependence of surface reflectivity on surface temperature. A more realistic analysis requires numerical calculations, which are described in Section 1.4.

1.4 Numerical simulation of laser-produced melting and evaporation

In the previous sections, we considered laser-produced melting and vaporization as separate and independent processes. Generally, these processes are coupled and should be considered jointly. This circumstance strongly complicates the analysis of the laser-induced phase transitions. Additional complications arise due to laser light reflection from the evaporation front. The reflectivity depends on the temperature and phase composition of the surface layer of the material. This layer is strongly nonuniform, and its reflectivity can be calculated using Maxwell's equations. It is clear that a quantitative consideration of this problem demands numerical calculations. We shall now discuss some of the results of such calculations.[10]

The model employed by Anisimov and Barsukov[10] includes the heat conduction equations for the temperatures of solid and liquid phases, with boundary conditions (1.14) and (1.15) on the melting front and (1.18) on the vaporization front. These equations can be written as:

$$\rho_s c_s \partial T_s/\partial t = \partial/\partial z(\kappa_s \partial T_s/\partial z) + Q_s(z,t) \quad Z_m(t) < z < \infty$$
$$\rho_l c_l \partial T_l/\partial t = \partial/\partial z(\kappa_l \partial T_l/\partial z) + Q_l(z,t) \quad Z_v(t) < z < Z_m(t)$$
(1.22)

The boundary conditions read:

$$\begin{aligned}
L_m \dot{Z}_m(t) &= \kappa_s \partial T_s/\partial z|_{z=Z_m} - \kappa_l \partial T_l/\partial z|_{z=Z_m} \\
L_v \dot{Z}_v(t) &= \kappa_l \partial T_l/\partial z|_{z=Z_m} \\
\dot{Z}_v(t) &= V_0 \exp(-U/T_v) \quad T_v = T_l(Z_v,t) \\
\dot{Z}_m(t) &= V_m c_s(T_f - T_m)/L_m \quad T_f = T_s(Z_m,t)
\end{aligned}$$
(1.23)

Here, the subscripts s and l refer to the solid and liquid phases, respectively. The laser energy released in the liquid is written as:

$$Q_l(z,t) = \sigma_l(z,t)|E(z,t)|^2 \qquad (1.24)$$

where $\sigma_l(z,t)$ is the local electrical conductivity of the liquid material (at laser frequency ω_0), and $E(z,t)$ is the local amplitude of the electric field in the laser beam, its value at the surface being related to the incident laser radiation intensity, I_0, by:

$$|E(Z_v,t)|^2 = 4\pi I_0/c_0$$

Here, c_0 is the speed of light. An equation similar to (1.24) also can be written for the solid phase. The electric field inside the metal is described by the wave equation:

$$\partial^2 E/\partial z^2 + k_0^2 \epsilon(z) E = 0 \qquad k_0 = \omega_0/c_0 \qquad (1.25)$$

which follows directly from Maxwell's equations.

Equations (1.22), (1.23), and (1.25) were solved numerically. The dielectric constant of the metal, ϵ, was considered dependent on the local temperature and phase state of the metal. The temperature dependence of ϵ was calculated using the Drude-Zener model and experimental data on the optical properties of metals.[22] Note that the experimentally measured reflectivity of some metals may significantly differ from the Drude-Zener reflectivity.[23] This difference, however, decreases at high temperatures and low photon energies (< 1 eV). The heat conduction equations (1.22) were solved using the front-capturing numerical scheme, which provides a precise calculation of melting and vaporization front positions. Some of the results of these calculations are presented in figures 1.2 through 1.6. The results are given for silver; the laser intensity equals 10^9 W/cm^2. The initial reflectivity of cold silver corresponds

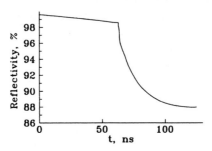

Figure 1.2: The reflectivity of a silver target as a function of time; $I = 10^8$ W/cm^2.

to a high-quality optical surface ($R \simeq 0.99$). In Figure 1.2, the change of surface reflectivity over time is shown. In the absence of melting, the reflectivity slowly decreases over time. When the liquid phase appears, the reflectivity changes drastically from its "solid" to "liquid" value; but there is no jump in reflectivity. Figures 1.3 and 1.4 show the temperature over time at the melting and evaporation fronts, respectively. Note that the melting front temperature is well above the melting point of silver, which is $T_m = 1234$ K. In Figure 1.5, the melting (curve 1) and vaporization (curve 2) front positions are shown as

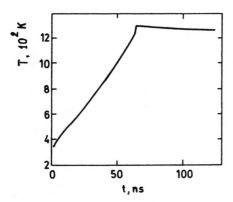

Figure 1.3: Temperature on the solid–liquid phase boundary as a function of time.

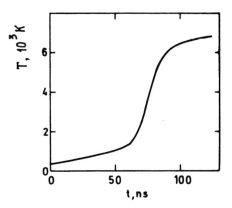

Figure 1.4: Temperature on the liquid–vapor phase boundary as a function of time.

functions of time. The melting front propagates in a stationary regime; but for the evaporation front, this regime has yet to be reached. Figure 1.6 shows the spatial profile of temperature in the vicinity of the vaporization front. As was mentioned in Section 1.3, the temperature has a maximum located inside the liquid phase, at a distance on the order of μ^{-1} from the phase boundary. Although the difference between the maximum temperature and the temperature at the phase boundary is rather small, it results in a qualitatively new phenomenon: the corrugation instability of the phase boundary.

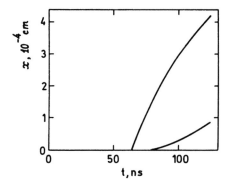

Figure 1.5: The positions of melting (curve 1) and vaporization (curve 2) fronts as functions of time.

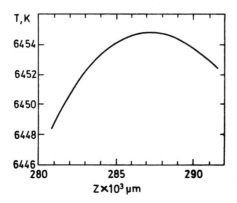

Figure 1.6: Spatial distribution of temperature in the liquid near the vaporization front.

REFERENCES

1. Ready, J. F., *Effects of High-Power Laser Radiation*, Academic Press, New York, 1971

2. Sokolov, A. V., *Optical Properties of Metals*, Blackie, London, 1967

3. Ziman, J. M., *Principles of the Theory of Solids*, Cambridge University Press, Cambridge, 1972

4. Libenson, M. N., Romanov, G. S., and Imas, Ya. A., Temperature dependence of the optical constants of a metal in heating by laser radiation, *Sov. Phys.-Tech. Phys.*, **13**, 925, 1968

5. Carslaw, H. S. and Jaeger, J. C., *Conduction of Heat in Solids*, Clarendon Press, Oxford, 1959

6. Prokhorov, A. M., Konov, V. I., Ursu, I., and Mihailescu, I. N., *Laser Heating of Metals*, Adam Hilger, Bristol, 1990

7. Shal'nikov, A. I., Institute for Physical Problems, Russian Acad. Sci., (private communication, 1967)

8. *Handbook of Physicochemical Properties of the Elements*, Ed. by Samsonov, G. V., IFI/Plenum Press, New York, 1968

9. Motorin, V. I. and Musher, S. L., Stability of liquefaction front in fast Joule heating, *Sov. Phys.-Tech. Phys.*, **27**, 726, 1982

10. Anisimov, S. I. and Barsukov, A. V., Phase transition kinetics and light reflection from metals heated by laser radiation, *Moscow Institute for Physics and Technology Report*, 1992

11. Landau, L. D. and Lifshitz, E. M., *Statistical Physics*, Pergamon Press, Oxford, 1980

12. Motorin, V. I. and Musher, S. L., Kinetics of volume melting. Nucleation and superheating of metals, *Journ. Chem. Phys.*, **81**, 465, 1984

13. Motorin, V. I., Absence of an elastic thermodynamic hysteresis in phase transitions in solids, *Sov. Phys.-Solid State*, **29**, 732, 1987

14. Anisimov, S. I., Imas Ya. A., Romanov, G. S., and Khodyko, Yu. V., *Action of High-Power Radiation on Metals*, National Technical Information Service, Springfield, Virginia, 1971

15. Frenkel, J., *Kinetic Theory of Liquids*, Dover Publishing, New York, 1955

16. Nesmeyanov, A. N., *Vapor Pressure of Chemical Elements*, Izd. AN SSSR, Moscow, 1961

17. Dushman, S., *Scientific Foundations of Vacuum Technique*, Wiley & Sons, New York, 1949

18. Anisimov, S. I., Bonch-Bruevich, A. M., Elyashevich, M. A., Imas, Ya. A., Pavlenko, N. A., and Romanov, G. S., Effect of powerful light fluxes on metals, *Sov. Phys.-Tech. Phys.*, **11**, 945, 1967

19. Anisimov, S. I., Evaporation of light-absorbing metals, *High Temperature*, **6**, 110, 1968

20. Anisimov, S. I., Dmitrenko, B. I., Leskov, L. V., and Savichev, V. V., Effect of surface reflectivity on the vaporization of a metal exposed to high-intensity light flux, *Fizika i Khimija Obrabotki Materialov* (Physics and Chemistry of Material Working, Russian) N 4, 10, 1972

21. Anisimov, S. I., Gol'berg, S. M., Sobol', E. N., and Tribel'skii, M. I., Oscillatory evaporation of condensed media by electromagnetic radiation, *Sov. Tech. Phys. Lett.*, **11**, 534, 1985

22. *Handbook of Optical Constants of Solids*, Ed. by Palik, E. D., Academic Press, New York, 1991

23. Huettner, B., Optical properties of polyvalent metals in the solid and liquid state: aluminium, *Journ. Phys., Condensed Matter* **B6**, 2459, 1994

Chapter 2

INTERACTION OF LASER RADIATION WITH DIELECTRICS

2.1 Optical breakdown of transparent dielectrics

In this chapter we will describe the effects that are produced when the laser beam is incident on transparent materials. Typical examples of materials under consideration are wide-band-gap dielectrics and semiconductors. When the laser intensity is below a certain threshold value, the material is transparent and the laser beam passes through the material with no apparent effect. But when the intensity is above the threshold, absorption abruptly increases, and effects such as melting, evaporation, and shock wave generation can occur. A violent increase in optical absorption of initially transparent material caused by a laser pulse is known as laser-induced breakdown. The laser breakdown is, in fact, a special kind of instability of a transparent dielectric in a high-intensity electromagnetic field.

Many different mechanisms of laser breakdown have been described in the literature (see references 1–3). The breakdown of real optical materials frequently is associated with the presence of inclusions, impurities, or structural imperfections of the material, in which the local absorption of light is much higher than the average absorption over the whole sample. In many cases of practical interest, the size of such absorbing inclusions is much less than the size of the focusing region.

In pure materials, on the other hand, the above extrinsic mechanisms of breakdown can be eliminated to a large extent. This implies a higher breakdown strength. The limiting strength is determined by intrinsic breakdown mechanisms, such as multiphoton ionization and electron avalanche.

Let us first consider the breakdown of transparent dielectrics initiated by absorbing inclusions.[4,5] There are two possible mechanisms of development of breakdown in this case. In the presence of inclusions of sufficiently large size, the most probable breakdown mechanism is local melting of material near the inclusions and formation of cracks as the result of thermal stresses. This mechanism of damage takes place, for example, in platinum-containing glasses.[6] It is possible, however, that there is another breakdown mechanism due to the fact that heating of the transparent medium in the vicinity of an absorbing inclusion leads to some additional absorption, which is equivalent to an increase in the size of the inclusion. Under certain conditions, this process results in the formation of a thermal wave propagating from the absorbing inhomogeneity through the volume of the medium. This wave arises as the result of thermal instability of the medium near the absorbing inclusion.

Consider the temperature field near an isolated absorbing inclusion. Let us assume for simplicity that the inclusion is spherical and is separated from the surrounding dielectric by a sharp boundary. In a quasi-stationary mode, the value of energy, which the inclusion acquires from the laser beam, is equal to the energy it transfers to the surrounding medium. We denote by I the laser intensity near the inclusion. We then can set the heat flux transferred from the inclusion to the surrounding dielectric equal to:

$$-\kappa \nabla T|_{r=R} = \alpha(R) I \qquad (2.1)$$

where κ is the thermal conductivity of dielectric, R is the radius of inclusion, and $\alpha(R)$ is the ratio of the absorption cross section of the inclusion to the area of its surface, $4\pi R^2$. The factor $\alpha(R)$ depends on the laser wavelength and can be calculated from the solution of the problem of light-wave diffraction at the inclusion (e.g., reference 7). In all cases of practical interest, $\alpha(R)$ is less than unity.

As mentioned above, the heating of the medium near the inclusion leads to some additional light absorption. We assume that a thermodynamic equilibrium is established near the inclusion so that the local absorption coefficient is governed by a local temperature. The temperature dependence of the absorption coefficient of media under consideration usually can be described by $\mu(T) = \mu_0 \exp(-E/T)$, where $2E = E_g$ is the forbidden band width. Thus, $\mu(T)$ is very strong function of temperature. The steady-state temperature distribution $T_s(r)$ can be found from the solution of the heat conduction equation:

$$\kappa \Delta T_s + I\mu(T_s) = 0 \qquad (2.2)$$

We will show that the steady-state temperature field, $T_s(r)$, turns out to be unstable if the laser intensity, I, exceeds a certain value. To analyze the stability of the solution of boundary-value problem given in Equations (2.2) and (2.1), we consider the transient heat-conduction problem and, as usual, assume that $T(r,t) = T_s(t) + \theta(r,t)$, $\theta(r,t) \ll T_s(r)$. For $\theta(r,t)$ we then find the linear equation:

$$c\rho \partial \theta / \partial t = \kappa \Delta \theta + I\mu'(T_s) \qquad (2.3)$$

where $\mu'(T_s) = d\mu(T_s)/dT_s$. We write the solution of Equation (2.3) as:

$$\theta(r,t) = \sum_\lambda C_\lambda \psi_\lambda(r) \exp(-\lambda t) \qquad (2.4)$$

where the functions $\psi_\lambda(r)$ satisfy the Schrödinger equation:

$$\Delta\psi + (\epsilon + V(r))\psi = 0, \quad V(r) = I\mu'(T_s)/\kappa, \quad \epsilon = c\rho\lambda/\kappa \qquad (2.5)$$

(the index λ is omitted from the function ψ for brevity). The boundary conditions for Equation (2.5) are:

$$\psi(\infty) = 0 \quad \text{and} \quad \partial\psi/\partial r|_{r=R} = 0$$

The second condition means that we consider temperature fluctuations in the medium while the thermal state of the inclusion remains constant. If among the solutions of Equation (2.5) there are those corresponding to negative values of λ, the steady-state temperature distribution, T_s, turns out to be unstable. By virtue of the previous arguments for the properties of the function $\mu(T)$, we conclude that the potential $V(r)$ in the Schrödinger equation (2.5) is large at $r \simeq R$ and decreases rapidly with increasing r. Thus, we are dealing with a standard quantum-mechanical problem of a bound state in a deep narrow well. Substituting $\psi = \varphi/r$ into (2.5) and integrating the resulting equation term by term, we find the instability condition:

$$\int_R^\infty \mu'(T_s(r))\,dr \geq \kappa/IR \qquad (2.6)$$

To evaluate the integral in (2.6), we transform the integration variable r to the new variable T_s and incorporate the fact that $\mu'(T_s)$ decreases rapidly with T_s decreasing. Then, the integration can be performed easily and yields:

$$IR\mu[T_s(R)] \geq \kappa |T_s'(R)|$$

Using the boundary condition (2.1), we arrive at the instability condition:

$$R\mu[T_s(R)] \geq \alpha \qquad (2.7)$$

The left side of Equation (2.7) is, in order of magnitude, the fraction of incident laser intensity, which is absorbed by heated dielectric material near the inclusion. When this fraction becomes larger than the fraction of incident laser intensity absorbed by the inclusion itself, the medium turns out to be unstable. In order to obtain the threshold intensity, one must solve the boundary value problem (2.1) and (2.2). The calculation carried out in reference 4 yields the critical laser intensity

$$I^* \approx \kappa E/R\alpha(R) \qquad (2.8)$$

Note that in the previous analysis, we assumed that the thermal conductivity of material does not depend on temperature. Although the breakdown criterion (2.7) is not associated with this assumption and is of a general nature, the value of the critical intensity (2.8) may be affected by the assumption made. The temperature dependence of thermal conductivity of the materials under consideration is due mainly to the electron component of heat conduction, which becomes the dominant energy transfer mechanism at high temperatures. The electron heat conduction plays a dual role in the breakdown process. On the one hand, it lowers the temperature, as well as the electron density near the surface of an inclusion; on the other hand, it increases the size of the absorbing halo around an inclusion and, consequently, enhances the absorbed energy. Therefore, estimating the effect of the electron heat conduction on the threshold of forming an absorbing wave is not a trivial task. The influence of the electron thermal conductivity in thermally ionized dielectric material on the characteristics of its optical breakdown initiated by an absorbing inclusion has been studied in reference 8. It was shown that allowing for electron thermal conductivity changes the critical intensity (2.8) relatively little.

As can be seen from Equation (2.8), the threshold of breakdown initiated by an individual inclusion depends substantially on its size (absorption cross-section). If the medium contains randomly distributed inclusions of different sizes, the breakdown threshold is no longer a completely defined quantity, and we can speak only on the probability of a breakdown under given conditions. Thus, the breakdown of a medium containing absorbing inhomogeneities should be described by statistical methods.

Let the distribution of laser intensity over the focal volume be $I(\mathbf{r}) = I_0 \varphi(\mathbf{r})$. To each intensity value there corresponds a critical size of inclusion $R^*(\mathbf{r})$, which is related to $I(\mathbf{r})$ by Equation (2.8). Breakdown inside the volume V will occur if the volume contains at least one absorbing inclusion whose size exceeds the critical value $R^*(\mathbf{r})$ at the corresponding point. Assuming that the distribution of inclusions is spatially uniform, it is easy to write an expression for the breakdown probability in the volume V:

$$P[I(\mathbf{r})] = [1 - \exp(-nV)]^{-1} \left\{ 1 - \exp\left[-n \int_{(V)} d^3r \int_{R^*(r)}^{\infty} f(\rho)\, d\rho \right] \right\} \quad (2.9)$$

Here n is the average density of absorbing inclusions; $f(R)$ is the distribution function of inclusions in size, normalized to unity; and $R^*(\mathbf{r})$ is the function related to $I(\mathbf{r})$ by Equation (2.8). In the experimental study of breakdown, the spatial profile of the intensity usually remains fixed, and the maximum intensity I_0 is varied. The breakdown probability in this case is a function of I_0, and using Equation (2.9), one can determine the average value and dispersion of the random value of the breakdown threshold. The calculations conducted in reference 5 show that, when the number of inclusions in the focal volume is large, $N = nV \gg 1$, the average value of breakdown intensity turns out to be strongly dependent on the focal volume V and the asymptotic behavior of the distribution function of inclusions $f(R)$ at large R. Assuming that as $R \to \infty$ the distribution function $f(R)$ falls out as $f(R) \propto R^{-k}$, the average breakdown intensity, $<I_0>$, was found to be proportional to $V^{-1/(k-1)}$. The last dependence is in satisfactory agreement with the experimental data[9] if one assumes the exponent $k = 2.5$. This "size effect" disappears when the focal volume and/or average concentration of inclusions is small, $N = nV \ll 1$. In this case, however, the breakdown usually is not due to absorbing inclusions in the medium, but to such intrinsic mechanisms as multiphoton ionization and electron avalanche.

The previous analysis of the instability of transparent dielectric materials containing absorbing inclusions was based on study of small temperature perturbations. This kind of linear analysis can yield only the instability criterion. To study the further development of instability, one should solve a much more complicated nonlinear problem. As indicated earlier, the consequence of the instability is the occurrence of an absorption wave. The propagation of this wave from the inclusion to the volume of the medium leads to the expansion of the ionized region inside the insulator that results finally in the absorption of a considerable portion of the laser energy and the appearance of macroscopic damage. In reference 8 numerical modeling has been employed to find the qualitative features of this phenomenon. It was assumed that the laser radiation impinges on a glass target containing a small metallic inclusion. The temperature field in the inclusion and glass matrix was described by the set of heat-conduction equations:

$$c_i \rho_i \partial T_i / \partial t = r^{-2} \partial / \partial r (r^2 \kappa_i \partial T_i / \partial r) + Q_i \qquad (2.10)$$

where $i = 1$ for the inclusion and $i = 2$ for the glass matrix. The set of equations (2.10) was solved numerically subject to the following initial and boundary conditions:

$$T_1(r,0) = T_2(r,0) = T_2(\infty,t) = T_0$$
$$\partial T_1/\partial r|_{r=0} = 0 \quad (\kappa_1 \partial T_1/\partial r - \kappa_2 \partial T_2/\partial r)|_{r=R} = 0 \qquad (2.11)$$

where R is the radius of inclusion. A small inclusion with a radius much smaller than the wavelength of laser radiation was considered. If we ignore the "shadow" thrown by the inclusion and suppose that the absorption in the halo is weak, we can assume that the temperature field has a spherical symmetry, as allowed by Equation (2.10). The dielectric in the vicinity of inclusion was assumed to be a partially ionized dense plasma. Transport and optical properties of this medium were expressed in terms of the effective collision frequency. The details of calculations of Q_i and κ_i are described in reference 8. We shall discuss here some of the results of the modeling. Figure 2.1 shows the spatial distribution of the temperature for different times. When the intensity is below the breakdown threshold defined by (2.8), the stationary temperature profile is established in the vicinity of inclusion (dashed curve). However, when the intensity exceeds the breakdown threshold, the absorption wave is formed, which propagates from the surface of the inclusion to the volume of dielectric (full curves). The size of the ionized halo as a function of time is shown in Figure 2.2. For laser intensities below the threshold, the size of the heated region saturates after some transient process (curve 1). For supercritical laser intensities, the front of thermal wave asymptotically approaches the regime with constant propagation velocity (curve 2). The

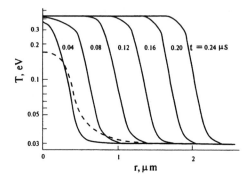

Figure 2.1: Temperature profiles at different times.
Dashed curve—steady-state temperature profile for $I = 2 \times 10^6$ W/cm$^2 < I^*$,
Solid curves—temperature in the thermal wave for $I = 2 \times 10^7$ W/cm$^2 > I^*$,
$I^* = 10^7$ W/cm^2.

glass matrix parameters used in calculations[8] were chosen to correspond to the experimental conditions.[10] The velocity of a steady-state absorption wave and the maximum temperature reached in this wave are in agreement with the experimental results.[10]

We considered the thermal ionization mechanism, in which absorption results from the appearance of free carriers in the conduction band of a solid. Other mechanisms of additional absorption may play roles in laser-induced breakdown of transparent dielectrics. There are a number of thermochemical processes resulting from the laser heating of a solid. They include the decomposition of complex compounds and formation of oxides. If the absorption coefficient of reaction products at the laser frequency is significantly high, these processes lead to thermochemical instability. This instability plays, for example, an important role in laser breakdown of polymers.[11] It was established

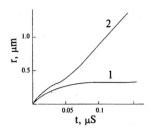

Figure 2.2: Position of the thermal wavefront as a function of time.
1—$I = 2 \times 10^6$ W/cm$^2 < I^*$, 2—$I = 2 \times 10^7$ W/cm$^2 > I^*$, $I^* = 10^7$ W/cm^2.

that the thermochemical mechanism involving pyrolysis and the formation of highly absorbing soot particles is responsible for the laser-induced damage in polymethylmethacrylate under the action of continuous laser radiation.[12,13] The qualitative picture of instability development and absorption halo formation is similar to that of the thermal ionization mechanism. However, the pyrolysis reaction requires a long time[13] and does not play a significant role in the case of short laser pulses.

Now, let us briefly consider the intrinsic mechanisms of breakdown of transparent dielectrics: multiphoton and avalanche ionization. The electron avalanche is the main mechanism of breakdown of gases, as well as solid and liquid dielectrics in both d.c. and high-frequency electric fields (see references 14–16). The avalanche is the most probable mechanism of laser breakdown in ultrapure transparent dielectrics, if the laser pulse length is not too short and the focal volume is not too small. As in the case of thermal ionization, the electrons in the conduction band acquire energy from the laser beam due to electron-phonon collisions. High-energy electrons then can produce the ionization, leading to electron multiplication. The major difference between the thermal and avalanche ionization is in the distribution function of electrons in the conduction band. In the case of avalanche, it is strongly nonequilibrium and should be found from the solution of the quantum kinetic equation.[16] As applied to laser breakdown of transparent dielectrics, different approximate solutions of this equation have been obtained in references 17–19. The distribution function of electrons usually is assumed to have the form:

$$f(\mathbf{p}, t) = f(\mathbf{p}) \exp(\gamma t)$$

where $\mathbf{p}$ is the momentum of the electron, and the avalanche development constant γ is calculated as a function of crystal parameters and laser pulse characteristics. The electron distribution function is supposed to be independent of spatial coordinates. When γ is determined, the threshold intensity is calculated from the breakdown criterion, which can be taken in the form: $\gamma \tau_l > C$, where C is either a constant or a weak (logarithmic) function of the crystal parameters and concentration of seed electrons. It generally is accepted that the concentration of seed electrons, n_0, is not too small, so that $n_0 V > 1$, where V is the focal volume. In the opposite case, $n_0 V < 1$, the breakdown threshold becomes a random quantity determined by the probability of appearance of a seed electron inside the focal volume.

Multiphoton ionization plays a dual role in the laser-induced breakdown of ultrapure insulators. On one hand, it is an auxiliary process responsible for supplying seed electrons necessary for the avalanche breakdown. On the other hand, the multiphoton ionization is a competitive mechanism of breakdown, which can provide, under certain conditions, a higher rate of generation of electrons in the conduction band than in the electron avalanche. The analysis of the relative contribution of the two processes was performed by Gorshkov et al.[20] It was shown that the dependence of the breakdown thresholds on the laser pulse length is different for avalanche and multiphoton breakdown mechanisms. Multiphoton ionization proved to be the dominant breakdown mechanism in picosecond and subpicosecond ranges of pulse lengths.

2.2 Stationary evaporation waves in transparent dielectrics

In Section 2.1 we described the propagation of a laser-supported absorption wave inside a transparent dielectric. The formation of the wave was initiated by an absorbing inclusion. In this section we shall consider the wave initiated by a "seed" located at the surface of transparent material. The experiments[21] show that in this case the propagation of absorption wave from the seed to the volume of dielectric is accompanied by the surface evaporation. Under certain conditions, synchronous propagation of the absorption and vaporization waves appeared to be possible. It also was shown that, for millisecond laser pulses, the stationary evaporation regime of the entrance and exit surfaces is established in silicate glasses and fused quartz.[21]

Consider a planar, steady-state evaporation wave propagating through the material with a strongly temperature-dependent absorption coefficient. Let the velocity of the wave be V_s. The temperature distribution in the wave is described by the heat conduction equation transformed into the moving frame, which is bound to the vaporization front:

$$c\rho V_s \partial T_s/\partial z + \partial/\partial z[\kappa(T_s)\partial T_s/\partial z] + Q(T_s) = 0 \qquad (2.12)$$

where $Q(T) = I(z)\mu(T)$. The equation for laser radiation intensity can be written as:

$$\partial I/\partial z = \mp \mu(T_s) I \qquad (2.13)$$

where minus and plus signs in the right-hand side correspond to the evaporation of the entrance and exit surfaces, respectively (see the schematic in Figure 2.3). The boundary conditions for Equations (2.12) and (2.13) are:

$$\kappa(T_s)\partial T_s/\partial z|_{z=0} = \rho V_s \Delta H \approx \rho V_s L_v \qquad T_s(\infty) = 0 \qquad (2.14)$$

$$I(0) = I_0 \qquad \text{(evaporation of the entrance surface)}$$

$$\text{or} \quad I(\infty) = I_0 \qquad \text{(evaporation of the exit surface)}$$

Note that only a part of the incident laser intensity is absorbed by the dielectric. The transmitted laser intensity ($I(\infty)$ in the case of entrance surface evaporation, and $I(0)$ in the case of exit surface evaporation), is not known *a priori* and should be determined in the course of the solution. As in Section 2.1, we assume the dependence $\mu(T)$ in the form $\mu(T) = \mu_0 \exp(-E/T)$ where $E = E_g/2$. The dependence of vaporization front velocity on the temperature is taken in the form of Equation (1.16): $V_s = V_0 \exp[-U/T_s(0)]$.

Substituting Equation (2.13) into Equation (2.12) we obtain the first integral, a form of the energy-flux conservation:

$$\kappa(T_s)\partial T_s/\partial z = \pm[I - I(\infty)] - c\rho V_s T_s \qquad (2.15)$$

If we set $z = 0$ in Equation (2.15) and employ the boundary conditions (2.14), we obtain the familiar formula:

$$\rho V_s(L_v + cT_s(0)) = \pm[I(0) - I(\infty)] = I_{abs} \qquad (2.16)$$

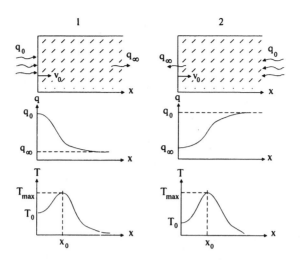

Figure 2.3: Schematic of the experiment for observing the evaporation of entrance (1) and exit (2) surfaces of a transparent dielectric. Qualitative profiles of the temperature and laser intensity are shown for both cases.

which coincides, formally, with (1.20) and can be combined with (1.16) to calculate the velocity V_s and surface temperature $T_s(0)$. In practice, however, this cannot be done since the absorbed laser intensity I_{abs} is not known.

The determination of temperature profile in the stationary evaporation wave is reduced to the integration of the coupled set of equations (2.13) and (2.15), subject to the boundary conditions

$$T_s(\infty) = 0, \qquad I(0) = I_0 \qquad (2.17)$$

(or $I(\infty) = I_0$ for exit surface evaporation). The velocity V_s is an eigenvalue of problem (2.13) and (2.15). It can be shown that, due to the invariance of the above problem under the transformation $z \to z + z_0, z_0 = $ constant, the boundary conditions (2.17) are sufficient to determine the solution $T_s(z)$ and eigenvalue V_s uniquely.

Equations (2.13) and (2.15) were solved numerically for fused quartz by Aleshin et al.[21] Figure 2.4 shows the dependence of the fraction of laser energy absorbed in the target, $A = I_{abs}/I_0$, on the incident light intensity I_0. Curves 1 and 2 correspond to the evaporation of incident and exit surface, respectively. The absorption decreases as the laser intensity increases in the range 10^6–10^7 W/cm^2 and remains approximately constant for intensities in the range 10^7–10^8 W/cm^2. The absorption is higher for the evaporation of exit surface than for the evaporation of entrance surface. At high intensities, when the absorbing layer becomes optically thin, the spatial profile of laser intensity is independent of the direction in which the radiation propagates, and the difference in the evaporation of the entrance and exit surfaces disappears. The stationary velocity of evaporation wave V_s is shown in Figure 2.5 as

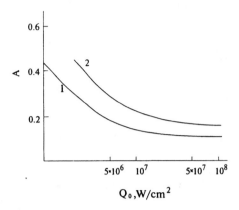

Figure 2.4: Absorption $A = I_{abs}/I_0$ of quartz glass as a function of incident laser intensity: 1—entrance surface evaporation, 2—exit surface evaporation.

a function of the incident light intensity. The velocity is proportional to the laser intensity in the range $I_0 > 10^7$ W/cm^2, where the absorption is constant.

The calculations show that in the case of exit surface evaporation, a stationary solution exists if the absorption is lower than a certain critical value $A < A^*$ (or the incident intensity is higher than some threshold value I^*). When $A > A^*$, the velocity of the absorption wave becomes higher than the velocity of the evaporation front. This results in the separation of the absorption layer from the phase boundary and in the dumping of vaporization. Thus we see that, under certain conditions, the laser evaporation of the exit surface

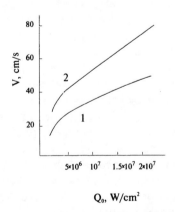

Figure 2.5: Stationary velocity of the evaporation wave as a function of incident laser intensity: 1—entrance surface evaporation, 2—exit surface evaporation.

of a transparent dielectric can be unstable. This instability will be studied in Chapter 6. In the case of entrance surface evaporation, a stationary solution exists for any intensity value of the incident radiation.

Since the absorption coefficient $\mu(T)$ is a strong function of temperature, the heat source in Equation (2.12) is localized in a thin layer near the maximum of temperature. This fact has been used for approximate solution of Equations (2.12) and (2.13).[22] The results obtained by Kovalev[22] are in a good agreement with the numerical solution[21] at fairly high laser intensities when the absorption layer is optically thin.

2.3 UV-laser ablation of polymers

Material removal caused by laser irradiation is often denoted as laser ablation. Recent experiments[23,24] show that for polymer materials the kinetics and dynamics of material removal strongly differ for UV and IR laser radiation. This means that the mechanism of laser-polymer interaction depends on the photon energy, $h\nu$. Thus, the traditional question of the competition of thermal and nonthermal mechanisms in laser ablation arises. A similar question also has been discussed in the laser annealing of semiconductors.[25] Recently, many experiments on UV-laser ablation of polymers have been performed with excimer lasers and pulse durations on the order of 10–30 ns. These experiments show that the products of ablation of organic polymers consist of low-molecular fragments (carbon, simple organic molecules) and medium-molecular fragments (monomers and parts of them).[26] To create such fragments, covalent bonds within the original polymer must be broken. If these bonds are broken due to thermal motion of atoms in molecules, the ablation velocity can be described as:

$$V = V_0 \exp(-E/T_0) \qquad (2.18)$$

where T_0 is the surface temperature and E is the bond-breaking energy. Typically, this energy is about 3–5 eV. Thus, experimentally observed ablation velocities of the order of 0.1 μm/pulse $\approx$ 1 m/s can be obtained only at very high temperatures. The theoretical model[27] yields temperatures about $T_0 \approx (6-12) \times 10^3$ K near the ablation threshold. These values contradict the experimental data on the vibrational temperatures of ablation products,[26] $T_{vib} \approx (2-3) \times 10^3$ K. The experiments also show that the UV-laser ablation of heat-sensitive polymers can be performed without damaging the remaining material, specifically with no trace of melting. The above facts cannot be explained on the basis of the thermal mechanism of ablation.

Photochemical ablation deals with direct breaking of chemical bonds as the result of light absorption. For the case of one-photon absorption this yields the dependence of the ablated layer thickness Δh on the laser fluence Φ in the form:

$$\Delta h = \mu^{-1} \log(\Phi/\Phi_{th}) \qquad (2.19)$$

where Φ_{th} is the threshold fluence for ablation. In recent experiments,[28] the dependence $h(\Phi)$ has been measured for polyimide with four different laser wavelengths. Only with the ArF-laser ($\lambda = 193$ nm) was agreement obtained with Equation (2.19) at $\Phi \approx \Phi_{th}$. All other wavelengths revealed dependencies that disagree with (2.19). Thus, the experimental results can neither be explained on the basis of a purely photochemical process, nor on the basis of a purely thermal process. They can, however, be described on the basis of a photophysical ablation mechanism.[29,30] The main idea of this mechanism is that the removal of electronically excited species from the surface requires a smaller activation energy than the removal of species in the electronic ground state.

Let us assume that the total ablation velocity is determined by both ground-state species A and excited-state species A^*. The excitation process $A \to A^*$ is induced by one-photon absorption. The dissipation of the excitation energy $A^* \to A$ occurs either via stimulated emission, or via nonradiative transitions. The nonradiative transitions will be characterized by a single thermal relaxation time τ_T, which is assumed to be independent of temperature. We assume also that a part of the excitation energy is lost via activated desorption of excited species.

Consider a planar ablation front. Neglecting any cooperative phenomena in the desorption process, the ablation front velocity can be described as:

$$V = xV_0^* \exp(-E^*/T_0) + (1-x)V_0 \exp(-E/T_0)$$

where $x = N^*/(N^* + N)$ is the normalized density of excited species; N and N^* are the number densities of ground-state and excited-state species, respectively; and E and E^* are their respective activation energies of desorption. We assume $E^* \ll E$ and, additionally, that E is on the order on the bond-breaking energy. In a coordinate frame connected with the ablation front, the density of species can be described by:

$$\begin{aligned} \partial N^*/\partial t &= V\partial N^*/\partial z + (\sigma I/h\nu)(N - N^*) - N^*/\tau_T \\ \partial N/\partial t &= V\partial N/\partial z - (\sigma I/h\nu)(N - N^*) + N^*/\tau_T \end{aligned} \quad (2.20)$$

where σ is the absorption cross section. Note that $N + N^* = N_0 =$ constant is the total number density of chromophores. According to Sauerbrey and Pettit,[31] it is typically $N_0 \simeq 6 \times 10^{21}$ cm^{-3}.

The propagation of the laser light is described by the equation:

$$\partial I/\partial z = -\sigma(N - N^*)I \quad (2.21)$$

The energy transfer in the solid polymer is described by the heat-conduction equation:

$$\partial T/\partial t = V\partial T/\partial z + \chi\partial^2 T/\partial z^2 + (\chi/\kappa)Q \quad (2.22)$$

where κ is the thermal conductivity, which is assumed to be temperature-independent, and χ is the thermal diffusivity. The heat source Q is due to the nonradiative transitions and is given by:

$$Q = h\nu N^*/\tau_T \quad (2.23)$$

CHAPTER 2. INTERACTION WITH DIELECTRICS

The boundary conditions at the ablation front $z = 0$ can be written as

$$\kappa \partial T/\partial z = \rho[x\Delta H^* V_0^* \exp(-E^*/T) + (1-x)\Delta H V_0 \exp(-E/T)]$$
$$I(0,t) = I_0 \qquad \text{at } z = 0 \qquad (2.24)$$

In addition to (2.24), we employ the boundary conditions

$$N(z,t) \to N_0, \ N^*(z,t) \to 0, \ T(z,t) \to 0 \quad \text{as } z \to \infty \qquad (2.25)$$

We now concentrate on stationary solutions of the boundary value problem (2.20), (2.21), (2.22), (2.24), and (2.25). If we set all time derivatives equal to zero in Equations (2.20)–(2.22) and assume that $I_0 =$ constant, we obtain the first integral

$$I_s = V_s(c\rho T_s + h\nu N_s^*) + \kappa \partial T_s/\partial z \qquad (2.26)$$

which represents the energy-flux conservation. Here, the subscript s indicates stationary quantities. Combining Equation (2.26) at $z = 0$ with boundary condition (2.24), we obtain the equation connecting the number density of excited-state species $N_s^*(0)$ with the surface temperature $T_0 = T_s(0)$; it can be written as:

$$Ax_0^2 + Bx_0 + C = 0 \qquad (2.27)$$

where
$$A = h\nu N_0 \left[V_0 \exp(-E/T_0) - V_0^* \exp(-E^*/T_0)\right]$$

$$B = (c\rho T_0/h\nu N_0)A + \rho\left[\Delta H V_0 \exp(-E/T_0) - \Delta H^* V_0^* \exp(-E^*/T_0)\right]$$

$$C = I_0 - (N_0 h\nu + c\rho T_0 + \rho\Delta H)V_0 \exp(-E/T_0)$$

and
$$x_0 = N_s^*(0)/N_0$$

With $\tau_T =$ constant, Equations (2.20) and (2.21) become independent from (2.22), and reduce to the single first-order differential equation[29]

$$V_s dN_s^*/dI_s = 1/h\nu - N^*/\tau_T \sigma I_s(N_0 - 2N_s^*) \qquad (2.28)$$

with the boundary condition: $N_s^* = 0$ at $I = 0$ (i.e., when $z \to \infty$). Integration of (2.28) from $I_s = 0$ to $I_s = I_0$ permits the calculation of $N_s^*(0) = x_0 N_0$. We then have the set of equations (2.27) and (2.28) to calculate the parameters of a stationary ablation wave: $N_s^*(0)$, T_0, and V_s. Numerical calculation of these parameters was performed in reference 29. Figure 2.6 shows the normalized ablation velocity $\tilde{V} = V_s h\nu N_0/I_b$ as a function of normalized laser intensity, I_0/I_b for different relaxation times τ_T. The value of constant I_b was chosen to be 10^7 W/cm^2. As seen from Figure 2.6, the ablation velocity increases as the relaxation time increases.

The dependence of the surface temperature on the laser intensity is plotted in Figure 2.7 for different values of τ_T. The full curves refer to the activation energy $E = 3$ eV and $E^* = 0.3$ eV. The dashed curves have been calculated for $E = 4.5$ eV. The temperature is given in the units $T_b = 4 \times 10^3$ K. We

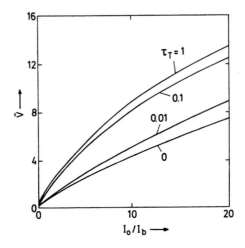

Figure 2.6: Normalized ablation velocity $\tilde{V} = V_s N_0 h\nu / I_b$ as a function of laser intensity at different relaxation times τ_T.
$I_b = 10^7$ W/cm², τ_T in units 10^{-8} s.

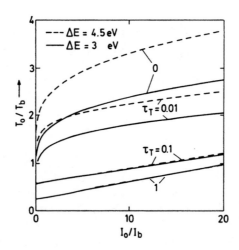

Figure 2.7: Ablation front temperature as a function of laser intensity for different τ_T.
$I_b = 10^7$ W/cm², $T_b = 4 \times 10^3$ K, τ_T—in units 10^{-8} s.

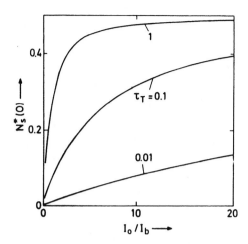

Figure 2.8: Concentration of excited species as a function of laser intensity. $I_b = 10^7$, τ_T—in units 10^{-8} s.

see that the surface temperature decreases as the result of energy relaxation. The "cold" ablation related to the deposition of excited species becomes increasingly important. With higher activation energies, the difference between thermal and photophysical ablation becomes more pronounced. In Figure 2.8 the concentration of excited species $N_s^*(0)$ is given as a function of laser intensity at different relaxation times. Note that N_s^* saturates at high laser intensities.

The spatial distribution of the normalized temperature $T_s(z)/T_0$, is shown in Figure 2.9 for different values of I_0 and τ_T. The coordinate z is given in units $z_0 = (\sigma N_0)^{-1}$. The width of the temperature distribution increases with intensity I_0 and relaxation time τ_T. On an expanded spatial scale, a maximum of temperature at some finite distance $z \ll z_0$ can be observed. This maximum is related to the finite penetration depth of the incident laser radiation. The spatial profile of $N_s^*(z)/N_0^*$ is shown in Figure 2.10 for different I_0 and τ_T. The concentration of excited species reaches its maximum at the surface $z = 0$.

The model described above explains many features observed in UV-laser ablation of organic polymers. With activation energies of $E = 3$ eV to 5 eV, $E^* = 0.3$ eV and relaxation times $\tau_T \simeq 1$ ns the model predicts realistic ablation rates at moderate surface temperatures of about 2000 K. The model also predicts that, although the stationary temperature has a maximum at a finite depth, the planar ablation wave is stable with respect to small perturbations.[30] This explains why pulsed UV-laser ablation provides smooth surfaces with polymers and rough surfaces with metals.

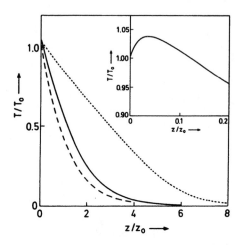

Figure 2.9: Spatial temperature profiles at different I_0 and τ_T.

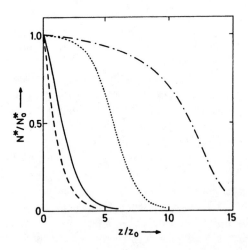

Figure 2.10: Spatial profiles of the concentration of excited species for different I_0 and τ_T.

REFERENCES

1. Ready, J. F., *Effects of High-Power Laser Radiation*, Academic Press, New York, 1971, Ch. 6

2. Manenkov, A. A. and Prokhorov, A. M., Laser-induced damage in solids, *Sov. Phys.-Uspekhi*, **29**, 104, 1986

3. Braunlich, P., Jones, S. C., Shen, X. A., Casper, R. I., Cartier E., DiMaria, D. J., and Fischetti, M. V., Non-avalanche dielectric breakdown in wide-band gap insulators at DC and optical frequencies, In: *Laser-Induced Damage in Optical Materials: 1989, Proceedings of the Boulder Damage Symposium 1989*, Ed. by Bennett, H. E., Chase, L. L., Guenther, A. H., Newnam, B. E., and Soileau, M. J., NIST Special Publ. 801, Washington, 1990

4. Anisimov, S. I. and Makshantsev, B. I., Role of absorbing inclusions in the optical breakdown of transparent media, *Sov. Phys.-Solid State*, **15**, 743, 1973

5. Aleshin, I. V., Anisimov, S. I., Bonch-Bruevich, A. M., Imas, Ya. A., and Komolov, V. L., Optical breakdown of transparent media containing microinhomogeneities, *Sov. Phys.-JETP*, **43**, 631, 1976

6. Pitts, J. H., Modeling laser damage caused by platinum inclusions in laser glass, in: *Laser Induced Damage in Optical Materials:1985* Natl. Bureau of Standards Special Publ. 746, Washington, 1988

7. Landau, L. D. and Lifshitz, E. M., *Electrodynamics of Continuous Media*, Pergamon Press, Oxford, 1984

8. Anisimov, S. I., Gal'burt, V. A., and Tribel'skii, M. I., Influence of the electron thermal conductivity on the threshold and dynamics of the breakdown of insulators containing microinclusions, *Sov. Journ. Quantum Electron.*, **11**, 1010, 1981

9. Ginzburg, V. L., *The Propagation of Electromagnetic Waves in Plasmas*, Pergamon Press, Oxford, 1970

10. Klochan, E. L., Popov, S. P., and Fedorov, G. M., Thermal instability caused in a transparent dielectric by a quasi-cw laser pulse, *Sov. Tech. Phys. Lett.*, **6**, 194, 1980;
    Zelikin, N. V., Kask, N. E., Radchenko, V. V., Fedorov, G. M., Fedorovich, O. V., and Chopornyak, D. B., Observation of absorption waves in a transparent dielectric, *Sov. Tech. Phys. Lett.*, **4**, 522, 1978

11. Liberman, M. A. and Tribelsky, M. I., The role of chemical reactions in the laser destruction of transparent polymers, *Sov. Phys.-JETP*, **47**, 99, 1978

12. Butenin, A. V. and Kogan, V. Ya., Nucleation and evolution of a thermochemical instability at an absorbing inclusion in polymethil methacrylate by a cw laser beam, *Sov. Phys.-Tech. Phys.*, **24**, 506, 1979

13. Gol'berg, S. M., Matyushin, G. A., Pilipetsky, N. F., Savanin, S. Yu., Sudarkin, A. N., and Tribelsky, M. I., Thermochemical instability of transparent media induced by an absorbing inclusion, *Appl. Phys.*, **B31**, 85, 1983

14. Raizer, Yu. P., Breakdown and heating of gases under the influence of a laser beam, *Sov. Phys.-Uspekhi*, **8**, 650, 1966

15. Sparks, M., Mills, D. J., Warren, R., Holstein, T., Maradudin, A. A., Sham, L. J., Loh, E., and King, D. F., Theory of electron-avalanche breakdown in solids, *Phys. Rev.*, **B24**, 3519, 1981

16. Mel'nikov, V. I., Quantum kinetic equation for electrons in a high-frequency field, *JETP Lett.*, **9**, 120, 1969

17. Rubinshtein, A. I. and Fain, V. M., Theory of avalanche ionization in transparent dielectrics under the action of a strong electromagnetic field, *Sov. Phys.-Solid State*, **15**, 332, 1973

18. Epifanov, A. S., Manenkov, A. A., and Prokhorov, A. M., Theory of avalanche ionization induced in transparent dielectrics by an electromagnetic field, *Sov. Phys.-JETP*, **43**, 377, 1976

19. Vlasov, R. A. and Zhvavyi, S. P., Optical avalanche breakdown in ionic crystals, *Journ. Appl. Spectroscopy*, **24**, 224, 1976

20. Gorshkov, A. S., Epifanov, A. S. and Manenkov, A. A., Avalanche ionization produced in solids by large radiation quanta and relative role of multiphoton ionization in laser-induced breakdown, *Sov. Phys.-JETP*, **49**, 309, 1979;
    Vlasov, R. A. and Zhvavyi, S. P., Role of two-photon absorption in optical avalanche breakdown, *Journ. Appl. Spectroscopy*, **36**, 62, 1982

21. Aleshin, I. V., Bonch-Bruevich, A. M., Imas, Ya. A., Libenson, M. N., Rubanova, G. M., and Salyadinov, V. S., Laser-induced evaporation of a nonlinearly absorbing dielectric, *Sov. Phys.-Tech. Phys.*, **22**, 1400, 1977

22. Kovalev, A. A., Laser evaporation of a transparent solid dielectric, *Sov. Phys.-Tech. Phys.*, **24**, 616, 1979

23. Srinivasan, R., Ablation of polymethil methacrylate films by pulsed (ns) ultraviolet and infrared (9.17 μm) lasers: a comparative study by ultrafast imaging, *Journ. Appl. Phys.*, **73**, 2743, 1993

24. Srinivasan, R., Ablation of polyimide (Kapton^{TM}) films by pulsed (ns) ultraviolet and infrared (9.17 μm) lasers. A comparative study, *Appl. Phys.*, **A56**, 417, 1993

25. Akhmanov, S. A., Emel'yanov, V. I., Koroteev, N. I., and Seminogov, V. N., Interaction of powerful laser radiation with the surfaces of semiconductors and metals: nonlinear optical effects and nonlinear optical diagnostics, *Sov. Phys.-Uspekhi*, **28**, 1084, 1989

26. Srinivasan, R. and Braren, B., Ultraviolet Laser Ablation of Organic Polymers, *Chem. Rev.*, **89**, 1303, 1989

27. Cain, S. R., Burns, F. C., Otis, C. E., and Braren, B., Photochemical description of polymer ablation: absorption behavior and degradation time scale, *Journ. Appl. Phys.*, **72**, 5172, 1992

28. Kueper, S., Brannon, J., and Brannon, K., Threshold behavior in polyimide photoablation: single-shot rate measurements and surface-temperature modeling, *Appl. Phys.*, **A56**, 43, 1993

29. Luk'yanchuk, B., Bityurin, N., Anisimov, S., and Baeuerle, D., The role of excited species in UV-laser materials ablation, Part I: Photophysical ablation of organic polymers, *Appl. Phys.*, **A57**, 367, 1993

30. Luk'yanchuk, B., Bityurin, N., Anisimov, S., and Baeuerle, D., The role of excited species in UV-laser materials ablation, Part II: The stability of the ablation front, *Appl. Phys.*, **A57**, 449, 1993

31. Sauerbrey, R. and Pettit, G. H., Theory of the etching of organic materials by ultraviolet laser pulses, *Appl. Phys. Lett.*, **55**, 421, 1989

Chapter 3

PROCESSES IN THE VAPOR PLUME

In this chapter we shall consider the processes that take place in the laser-produced vapor cloud. We shall study the structure of a vapor flow which is important for the analysis of the interaction of incident laser radiation with the evaporated material. We also shall discuss the effect of the ambient gaseous atmosphere on vapor-cloud dynamics, including the instability of the contact boundary between the expanding vapor and ambient gas.

3.1 Hydrodynamic boundary conditions for strong evaporation

Consider the steady-state evaporation of a solid (liquid) into a vacuum. Let the temperature of the surface of the condensed phase be T_s. Let us choose a coordinate frame connected with the vaporization front with z-axis directed from the condensed phase to the gaseous one. It generally is accepted (references 1 and 2) that the atoms emitted from the surface of the condensed phase have the Maxwellian velocity distribution for $v_z > 0$, with number density equal to the equilibrium number density of the saturated vapor $n_s(T_s)$. For $v_z < 0$, a good approximation for the distribution function is $f \approx 0$. We therefore see that the velocity distribution of atoms near the surface is substantially different from the local equilibrium distribution. Consequently, in the immediate vicinity of the vaporizing surface, there is a layer of several mean free paths length, in which the distribution approaches equilibrium. This so-called Knudsen layer must be considered a discontinuity in the hydrodynamic treatment. The kinetic equation must be solved to determine the structure of this region and the values of hydrodynamic variables beyond the discontinuity. This analysis was conducted by Anisimov[3].

The problem under consideration resembles that of a strong shock wave (or rather a "strong rarefaction wave"): just as in a strong shock, a considerable change of the distribution function occurs, on the scale of several mean free paths. We can use, therefore, the approach proposed by Tamm[4] and Mott-Smith[5] in their studies of strong shock-wave structure. The special feature of this method is an approximation of the distribution function within the discontinuity region by the sum of distribution functions before and after the discontinuity with coordinate-dependent coefficients. This approach is acceptable if the region of the main change of distribution function is narrow.

The distribution function in the Knudsen layer can be written in the form:

$$f(z, \mathbf{v}) = \alpha(z) f_1(\mathbf{v}) + [1 - \alpha(z)] f_2(\mathbf{v}) \tag{3.1}$$

where

$$f_1(\mathbf{v}) = \begin{cases} n_s(T_s)(M/2\pi k T_s)^{3/2} \exp(-Mv^2/2kT_s) & v_z > 0 \\ \beta f_2(\mathbf{v}) & v_z < 0 \end{cases}$$

$$f_2(\mathbf{v}) = n_1(M/2\pi k T_1)^{3/2} \exp\{-(M/2kT_1)[(v_z - u_1)^2 + v_x^2 + v_y^2]\}$$

Here $\alpha(z)$ is the unknown function that satisfies the conditions $\alpha(0) = 1$ and $\alpha(\infty) = 0$, T_s is the surface temperature, and $n_s(T_s)$ is the saturated vapor density at this temperature. The values of n_1, u_1, and T_1 refer to an equilibrium state beyond the discontinuity and are related in the steady-state evaporation regime by the Jouguet condition:[3]

$$u_1 = c_s(T_1) \tag{3.2}$$

where $c_s(T) = (\gamma k_B T/M)^{1/2}$ is the speed of sound. The last condition follows from the fact that the expansion of vapor into a vacuum is described by the centered rarefaction wave.[6]

The conservation laws of mass, momentum, and energy fluxes hold within the discontinuity region:

$$\begin{aligned} \int d\mathbf{v}\, v_z f(z, \mathbf{v}) &= C_1 \\ \int d\mathbf{v}\, v_z^2 f(z, \mathbf{v}) &= C_2 \\ \int d\mathbf{v}\, v_z v^2 f(z, V) &= C_3 \end{aligned} \tag{3.3}$$

Equations (3.2) and (3.3) form the set of coupled equations for the calculation of flow parameters at the sonic point and the constant β. Solving these equation, we obtain:[3]

$$\beta = 6.29, \quad T_1 = 0.67 T_s, \quad n_1 = 0.31 n_s(T_s) \tag{3.4}$$

Note that for the adiabatic index γ, the value 5/3 was assumed, corresponding to a monatomic gas, since the equilibrium of internal degrees of freedom is established at a distance from the phase boundary that is much larger than the mean free path of atoms.

It can be shown using Equations (3.4) that the flux j_- of atoms condensed back on the phase boundary amounts to approximately 0.18 of the flux of evaporated atoms $j_+ = (k_B T_s/2\pi M)^{1/2} n_s(T_s)$, i.e., $j_- \ll j_+$. We now can obtain

CHAPTER 3. PROCESSES IN THE VAPOR PLUME

more precise formulas than those derived in Chapter 1 for the vaporization front velocity, vapor temperature, condensed phase temperature, and other parameters. Below are some results presented without detailed calculations. The surface temperature $T_s = T_s(0)$ in the stationary evaporation wave is determined by the equation (see Equations (1.20) and (1.21) for comparison):

$$n_s(T_s)(Mk_BT_s)^{1/2}(L_v + 2.2k_BT_s/M) = 3.1 I_{abs}$$

where L_v is the latent heat of vaporization per unit mass and I_{abs} is the laser intensity absorbed. The velocity of the vaporization front is determined by the formula:

$$V_s = \frac{I_{abs}}{\rho(L_v + 2.2k_BT_s/M)}$$

These results differ slightly from those derived without considering the vapor expansion dynamics. It is important to note that under certain conditions, the vapor at the sonic point appears to be supersaturated. This means that the condensation may begin very near the evaporating surface. To estimate the condition for the supersaturation to occur, we compare the vapor density at the sonic point $n_1 = 0.31 n_s(T_s)$ with the saturated vapor density corresponding to the temperature $T_1 = 0.67 T_s$. Assuming the exponential temperature dependence of saturated vapor density, we obtain the condition: $T_s < 0.4 M L_v/k_B$. We thus see that at moderate laser intensities, when the surface temperature is considerably lower than $U = ML_v/k_B$, the vapor near the surface is expected to be in the supersaturated state.

The foregoing considered the expansion of vapor into vacuum, but the same method of derivation of hydrodynamic boundary conditions can be employed in the case of expansion of the vapor into an ambient gaseous atmosphere. In this case the hydrodynamic velocity of vapor at the outer boundary of the Knudsen layer is determined by the external flow and is no longer equal to the local sound velocity. The jumps of density and temperature across the Knudsen layer become functions of the Mach number M_1 of the external flow at the outer boundary of the Knudsen layer. To determine the boundary conditions, we have to modify Equation (3.2) as follows:

$$u_1 = M_1 c_s(T_1) \tag{3.5}$$

and solve the set of equations (3.3) and (3.5). The dependences $T_1(M_1)$ and $n_1(M_1)$ resulting from the numerical solution of (3.3) and (3.5), can be approximated by the following analytical formulas:

$$T_1 \approx T_s(1 - 0.33 M_1), \qquad n_1 \approx n_s(T_s)/(1 + 2.2 M_1)$$

Again, we can compare the vapor density n_1 on the outer boundary of the Knudsen layer with the saturated vapor density corresponding to the temperature T_1. The comparison shows that the condition required for supersaturation to occur depends on the Mach number M_1. For example, at $M_1 = 0.5$ we have the condition $T_s < 0.2 L_v M/k_B$, and at $M_1 = 0.25$ $T_s < 0.16 L_v M/k_B$. Thus, the range of surface temperatures where supersaturation occurs decreases as M_1 decreases.

The above hydrodynamic boundary conditions are similar (if we extend the analogy with the problem of the strong shock wave) to the Hugoniot relations. The structure of vapor flow in the Knudsen layer (similar to the internal structure of a shock wave) can be found from the solution of the Boltzmann kinetic equation. Numerical methods have been employed widely to solve this equation for both stationary (references 7–9 and references therein) and nonstationary[10] evaporation. We will not discuss these works here. Notice only that the hydrodynamic boundary conditions resulting from the numerical solution of the Boltzmann equation are in close agreement with the previous simple analysis.

3.2 Condensation in expanding vapor

It was shown in the previous section that, under certain conditions, the vapor on the outer boundary of the Knudsen layer is in a supersaturated state. This state is unstable, so the condensation of the vapor will take place immediately beyond the Knudsen layer.[3,11] In typical experimental conditions ($T_s \ll U$), the supersaturation is high; the condensation therefore occurs very rapidly in a very narrow region, which can be regarded as a surface of discontinuity (condensation shock) separating the original vapor from a vapor containing droplets of condensed phase.[6] The vapor parameters before and after the discontinuity are related by the laws of conservation of mass, momentum, and energy fluxes. For one-dimensional flow, they read:

$$\begin{aligned} \rho_1 u_1 &= \rho_2 u_2 \\ p_1 + \rho_1 u_1^2 &= p_2 + \rho_2 u_2^2 \\ H_1 + u_1^2/2 &= H_2 + u_2^2/2 \end{aligned} \quad (3.6)$$

Here subscripts 1 and 2 refer to the states on the outer boundary of the Knudsen layer and on the outer boundary of condensation shock, respectively. The density of equilibrium two-phase system is $\rho = Mn$, $n = n_c + n_s$, where n_s is the saturated vapor number density and n_c is the number of atoms in condensed phase per unit volume of the two-phase system. Introducing the degree of condensation, $\alpha = n_c/n$, we can write the pressure and internal energy of the two-phase system in the following form:

$$p = p_s(T) = n(1-\alpha)k_B T, \quad \epsilon = (1-\alpha)L_v + 3(1+\alpha)k_B T/2M \quad (3.7)$$

Here we assumed that the heat capacity of the condensed phase is $3k_B$ per atom. Using the condition for the adiabaticity $dS = d\epsilon + pd(1/\rho) = 0$ and the expression for the saturated vapor pressure:[2]

$$p_s(T) = B_0 T^{-1/2} \exp(-U/T)$$

we can obtain the equation for the adiabatic of the saturated vapor in the form:

$$(y - 1/2)d\alpha + (3/y + \alpha - 1)dy = 0$$

where $y = U/T$. The solution of this equation, subject to the initial condition $\alpha = \alpha_2$ when $y = y_2$, is of the form:

$$(1 - \alpha)(y - 1/2) = (1 - \alpha_2)(y_2 - 1/2) + 3\log(y/y_2) \tag{3.8}$$

Using Equations (3.7) and (3.8), we obtain, after some straightforward calculations, a formula for the sound speed in two-phase mixture:

$$c_s = (\partial p/\partial \rho)_s^{1/2} = \frac{(k_B T/M)^{1/2}(1 - \alpha)(y - 1/2)}{[(y^2 - 3y + 3/4)(1 - \alpha) + 3]^{1/2}} \tag{3.9}$$

Since the above analysis is justified when $T \ll ML_v/k_B$, Equation (3.9) can be approximately reduced to $c_s \approx [(1 - \alpha)k_B T/M]^{1/2}$. Finally, using Equation (3.7) we can rewrite the third equation (3.6) in the form:

$$\begin{aligned} &5k_B T_1/2M + L_v + u_1^2/2 \\ &= (1 - \alpha_2)(5k_B T_2/2M + L_v) + 3\alpha_2 k_B T_2/M + u_2^2/2 \end{aligned} \tag{3.10}$$

In the presence of condensation shock, the Jouguet condition (3.2) should be set on the outer boundary of the condensation shock:

$$u_2 = c_s(T_2, \alpha_2) \tag{3.11}$$

where c_s is given by Equation (3.9).

Let us turn now to Equations (3.6). According to the results of Section 3.1, we can express the left sides of these equations as functions of a single parameter, u_1. Therefore Equations (3.6), supplemented by the Jouguet condition in the form (3.11), constitute a set of equations to determine three unknown parameters: T_2, α_2, and u_1. To simplify the calculations, we suppose that $T_s \ll ML_v/k_B$. In this case, as one can see from (3.10), α_2 is of the order of $\alpha_s \propto k_B T_s/ML_v \ll 1$, so we can neglect α_2 in the first two equations (3.6) and then use Equation (3.10) to calculate α_2. After simple calculations, we obtain the hydrodynamic boundary conditions on the outer boundary of condensation discontinuity, in the following form:

$$\begin{aligned} T_2/T_s &\approx 1 & n_2/n_s(T_s) &\approx 0.28 \\ u_2 &\approx (k_B T_s/M) & \alpha_2 &\approx 0.65 k_B T_s/ML_v \end{aligned} \tag{3.12}$$

3.3 Dynamics of vapor expansion: one-dimensional flow

The motion of vapor beyond the condensation shock obeys the gas-dynamics equations. Since the vapor is transparent, its motion is isentropic and is described by the self-similar solution of gas-dynamics equations.[6] For the case of one-dimensional expansion of a vapor, this solution depends on the similarity variable $\xi = z/t$, where the z-axis is directed perpendicularly to the phase boundary. Introducing the similarity variable into the equations of gas dynamics, we obtain:[6]

$$\begin{aligned}(u-\xi)\partial\rho/\partial\xi \;+\; \rho\partial u/\partial\xi \;=\; 0 \\ (u-\xi)\rho\partial u/\partial\xi \;+\; c_s^2\partial\rho/\partial\xi \;=\; 0\end{aligned} \quad (3.13)$$

One must add to the set of equations (3.13) a condition for the adiabaticity of an equilibrium two-phase flow. We write this condition in the form (3.8). From Equations (3.13) we obtain, after standard transformations:[6,11]

$$u = \xi + c_s = u_2 + \int_\rho^{\rho_2} c_s\, d\rho/\rho \quad (3.14)$$

Using Equations (3.8) and (3.9), we obtain from Equation (3.14) in the limiting case $k_B T_s/M \ll L_v$:

$$\begin{aligned}u(\xi)/c_2 &\approx 1 - (U/T_2)\log[T(\xi)/T_2] \\ &\approx \xi/c_2 + T(\xi)/T_2\end{aligned} \quad (3.15)$$

Equations (3.15) define the velocity and temperature of vapor as functions of the similarity variable ξ. We see that the dependence of the velocity on ξ is approximately linear, and the temperature decreases exponentially with ξ increasing.

It should be noted that the solution (3.15) is valid only in a limited range of ξ. Adiabatic expansion of the two-phase mixture results in a very fast drop in vapor density and, hence, in the "freezing" of condensation.[12] At some density, the degree of condensation α reaches its maximum value α_{max} and remains constant during subsequent expansion. A more detailed analysis of the freezing of condensation is described in reference 11. It was shown that, under certain conditions, the freezing can occur in the immediate vicinity of the phase boundary. The vapor expansion is described in this case by the simple ideal-gas relations given in references 6 and 12.

We have considered the expansion of vapor into vacuum. This consideration remains approximately valid when the pressure of ambient gas is much smaller than the saturated vapor pressure near the vaporization front. If, however, these two pressures are not very different, the effect of ambient atmosphere must be taken into account. The structure of one-dimensional flow in this case is shown schematically in Figure 3.1. The expanding vapor successively passes through a (1) Knudsen layer, in which the velocity distribution of evaporated atoms acquires the equilibrium Maxwellian form; (2) a condensation discontinuity, in which the supersaturated vapor is partially condensed; (3) a rarefaction wave; and, (4) a uniform flow region. The outer boundary of this region is (5) the contact surface, which separates the evaporated material and the gaseous ambient. The evaporated material acts on the surrounding gas much as a piston generating (7) a shock wave in the gas. The shock-compressed gas occupies (6) a uniform flow region, between the shock wave and contact surface. The solutions of gas-dynamics equations describing the

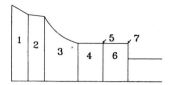

Figure 3.1: Expansion of vapor into a gaseous atmosphere. Qualitative structure of flow. 1—Knudsen layer; 2—condensation discontinuity; 3—rarefaction wave; 4—region of uniform vapor flow; 5—contact boundary; 6—uniform flow of shock-compressed gas; 7—shock wave.

flow characteristics in regions (3), (4), and (6) are well known.[6,12] These solutions should satisfy the boundary conditions on the discontinuity surfaces (2), (5), and (7). The boundary conditions on the shock wave follow from the conservation of mass, momentum, and energy fluxes across the shock front. These conditions can be written as:[6]

$$u_6 = 2D(1 - M_0^{-2})/(\gamma + 1), \quad p_6 = p_a \left[1 + 2\gamma(M_0^2 - 1)/(\gamma + 1)\right] \quad (3.16)$$

where D is the shock wave velocity, $M_0 = D/c_a$ is the Mach number of the shock wave, c_a is the sound speed in the ambient gas, p_a is the external pressure and γ is the adiabatic exponent of the gas. The conditions on the contact boundary (5) are:

$$u_6 = u_4 \quad \text{and} \quad p_6 = p_4 \quad (3.17)$$

Combining the boundary conditions discussed above on the condensation shock with Equations (3.14), (3.16), and (3.17), one can determine all characteristics of gas and vapor flows as functions of the surface temperature T_s and external pressure p_a. Numerical calculations for aluminum have been carried out by Igoshin and Kurochkin.[13] On the basis of these calculations, approximate analytical equations were proposed for making numerical estimates of the flow parameters.

3.4 Dynamics of vapor expansion: spatial structure of the vapor plume

Analysis of vapor expansion dynamics using one-dimensional approximation is justified when the thickness of the vapor layer $\Delta z \propto c_s \tau_l$ is much smaller than the laser spot size. This condition usually is satisfied for nanosecond laser pulses. The formation of the vapor/plasma cloud near the surface of a target can be described in terms of the one-dimensional model. However, the subsequent expansion of the cloud into a vacuum or ambient gas is a two- or three-dimensional process. The dynamics of this process have been studied experimentally for different laser sources and target materials.[14–18]

In this section we will consider a simple analytical model that yields a correct qualitative explanation of the basic experimental results. The model is based on a particular solution of the gas dynamics equations that results from the invariance of these equations under a Lie group transformation.[19] We suppose that the initial vapor cloud formed near the target surface has an ellipsoidal form (with semiaxes X_0, Y_0, and Z_0) and consider its expansion into vacuum. The expansion of the cloud is described by the gas dynamics equations:

$$\begin{aligned}\partial \rho/\partial t &+ \text{div}\,(\rho \mathbf{u}) = 0 \\ \partial \mathbf{u}/\partial t &+ (\mathbf{u}\nabla)\mathbf{u} + (1/\rho)\nabla p = 0 \\ \partial S/\partial t &+ \mathbf{u}\nabla S = 0\end{aligned} \qquad (3.18)$$

We assume that the vapor can be described by the ideal gas equation of state with the adiabatic exponent γ.

Consider a fluid motion in which the coordinates of a fluid particle $r_i(t)$, ($i = x, y, z$), change according to the equation:[20]

$$r_i(t) = \sum_k F_{ik}(t) r_k(0) \qquad (3.19)$$

where $r_k(0)$ are the initial coordinates. Transformation (3.19) describes uniform deformations and rotations of a fluid mass. If we ignore the rotation, the matrix $F_{ik}(t)$ is reduced to a diagonal form:

$$F_{ik}(t) = \begin{vmatrix} X(t)/X_0 & 0 & 0 \\ 0 & Y(t)/Y_0 & 0 \\ 0 & 0 & Z(t)/Z_0 \end{vmatrix} \qquad (3.20)$$

Equation (3.19) with the matrix F_{ik} given by (3.20) describes the conversion of fluid ellipsoid with semiaxes X_0, Y_0, and Z_0 into another ellipsoid, with semiaxes X, Y, and Z. According to Equation (3.19), the fluid velocity at the point (x, y, z) is proportional to the radius vector of this point. We have, then:

$$u_x = x\dot{X}/X \qquad u_y = y\dot{Y}/Y \qquad u_z = z\dot{Z}/Z \qquad (3.21)$$

where $\dot{X} = dX/dt$, $\dot{Y} = dY/dt$, $\dot{Z} = dZ/dt$. Substituting (3.21) into (3.18), we find that the density and pressure profiles in the vapor cloud should be in the following form:

$$\begin{aligned}\rho(x,y,z,t) &= \frac{M}{I_1 XYZ}\psi(x,y,z,t) \\ p(x,y,z,t) &= \frac{E}{I_2 XYZ}(X_0 Y_0 Z_0/XYZ)^{(\gamma-1)}[\psi(x,y,z,t)]^\gamma \\ \psi(x,y,z,t) &= \left[1 - \frac{x^2}{X^2} - \frac{y^2}{Y^2} - \frac{z^2}{Z^2}\right]^{1/(\gamma-1)}\end{aligned} \qquad (3.22)$$

where

$$M = \int \rho(x,y,z,t)\,dV \quad \text{and} \quad E = \int p(x,y,z,0)\,dV$$

CHAPTER 3. PROCESSES IN THE VAPOR PLUME

The constants M and E are proportional to the total mass evaporated and total initial energy of the gas cloud, respectively. The normalization constants, I_1 and I_2, are given by

$$I_1 = \pi\Gamma\left[\gamma/(\gamma-1)\right]\Gamma(3/2)/\Gamma\left[\gamma/(\gamma-1)+3/2\right]$$
$$I_2 = \pi\Gamma\left[(2\gamma-1)/(\gamma-1)\right]\Gamma(3/2)/(\gamma-1)\Gamma\left[\gamma/(\gamma-1)+5/2\right]$$

where $\Gamma(z)$ is the gamma-function.[21]

Substitution of (3.21) and (3.22) into the gas-dynamics equations (3.18) gives the following equations for the elements of the matrix (3.20):

$$d^2X/dt^2 = -\partial U/\partial X \quad d^2Y/dt^2 = -\partial U/\partial Y \quad d^2Z/dt^2 = -\partial U/\partial Z \quad (3.23)$$

Equations (3.23) are formally identical with the equations of motion of a particle in the classical mechanics, where the potential energy is

$$U(X,Y,Z) = \frac{\beta(\gamma)}{(\gamma-1)}\left[\frac{X_0 Y_0 Z_0}{XYZ}\right]^{\gamma-1} \quad (3.24)$$

$$\beta(\gamma) = 2\gamma I_1 E/(\gamma-1)I_2 M = (5\gamma-3)E/M$$

Initial conditions for the set of equations (3.23) can be written as $X(0) = X_0$, $dX/dt(t=0) = 0$ and similarly for Y and Z. We assume here that the initial kinetic energy of the vapor is much smaller than its thermal energy. Introducing dimensionless variables

$$\xi = X/X_0, \ \eta = Y/Y_0, \ \zeta = Z/Z_0, \ \eta_0 = Y_0/X_0, \ \zeta_0 = Z_0/X_0, \ \tau = t/t_0$$

where $t_0 = X_0 \beta^{-1/2}$, and X_0 is the largest axis of the initial ellipsoid, we transform Equation (3.23) as:

$$\xi\xi'' = \eta\eta'' = \zeta\zeta'' = (\eta_0\zeta_0/\xi\eta\zeta)^{\gamma-1} \quad (3.25)$$

$$\xi(0) = 1, \quad \eta(0) = \eta_0, \quad \zeta(0) = \zeta_0, \quad \xi'(0) = \eta'(0) = \zeta'(0) = 0$$

where $\xi' = d\xi/d\tau$, etc. Thus, the evolution of the vapor cloud in variables ξ, η, and ζ is determined by three parameters: γ, η_0, and ζ_0.

Equations (3.25) have the first integral (in terms of classical mechanics, the energy integral). A standard procedure leads to the following relation:

where
$$\xi'^2 + \eta'^2 + \zeta'^2 + 2U = C \quad (3.26)$$

$$U(\xi,\eta,\zeta) = (\gamma-1)^{-1}\left[\frac{\eta_0\zeta_0}{\xi\eta\zeta}\right]^{\gamma-1}$$

and $C = 2/(\gamma-1)$. For the special case $\gamma = 5/3$, there exists an additional integral of (3.25). This integral has been derived in reference 22 and it can be written as

$$\xi^2 + \eta^2 + \zeta^2 = 3\tau^2 + 1 + \eta_0^2 + \zeta_0^2 \quad (3.27)$$

Using (3.25) and (3.27), it can be shown readily that $\xi(\tau)$, $\eta(\tau)$, and $\zeta(\tau)$ are all linear functions of τ when $\tau \to \infty$. This means that the expansion of the vapor plume becomes inertial as the pressure gradients tend to zero. In general, Equations (3.25) have to be solved numerically. Numerical integration was performed in references 23 and 24 for a wide range of parameters η_0, ζ_0, and γ. The limiting shape of the expanding ellipsoid has been calculated for each set of parameters η_0, ζ_0, γ. It was shown by Anisimov et al.[22-24] that the maximum expansion velocity is reached in the direction of maximum initial pressure gradient. It means, in particular, that a disk-shaped vapor cloud is transformed into a cigar-shaped one and *vice versa*. It follows from the numerical calculation that the limiting expansion velocity in the z-direction is determined mainly by the parameter ζ_0 and does not really depend on the parameter η_0. In turn, the expansion in the y-direction is governed primarily by the parameter η_0. As seen in Figure 3.2, the change of ζ_0 by factor of 30 results only in a few percent change of the ratio Y/X. It appears, then, that the plume expansion can be considered as a superposition of two nearly independent motions. Both motions become inertial at $\tau > 10^3$–10^4. At later times, the ratios Y/X and Z/X remain close to their asymptotic values at $\tau \to \infty$. Some of these asymptotic values are presented in Table 3.4 for $\gamma = 7/5$.

ζ_0 \ η_0	0.7	0.5	0.3
0.01	1.219592	1.472280	1.965698
	5.486703	5.821500	6.461423
0.03	1.220410	1.473599	1.965206
	3.959977	4.178862	4.584814
0.1	1.217673	1.464305	1.930009
	2.653460	2.767006	2.961508
0.3	1.205422	1.430687	1.834970
	1.732922	1.773907	1.834970

Table 3.1: Values Y/X (upper lines) and Z/X (lower lines) at $t \to \infty$.

We can see in Table 3.4 that the expansion of the plume in the z-direction is essentially independent of its expansion in the (x, y)-plane.

3.5 Instability of the vapor-plume expansion into a vacuum

The solution obtained in the previous section is based on the assumption that the entropy (per unit mass) is constant over the initial vapor cloud. This assumption is physically reasonable. It is clear, however, that the real entropy profile depends on the details of the vaporization process and might be, generally, nonuniform. We will show in this section that the stability of the vapor plume expansion into a vacuum depends on the initial entropy profile.

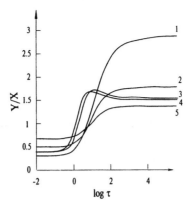

Figure 3.2: Evolution of the shape of a vapor cloud during its expansion into vacuum. The ratio Y/X is shown as a function of τ for different values of η_0, ζ_0, and γ.
1—$\eta_0 = 0.3$, $\zeta_0 = 0.01$, $\gamma = 5/3$,   2—$\eta_0 = 0.5$, $\zeta_0 = 0.01$, $\gamma = 5/3$,
3—$\eta_0 = 0.3$, $\zeta_0 = 0.3$, $\gamma = 1.2$,   4—$\eta_0 = 0.3$, $\zeta_0 = 0.01$, $\gamma = 1.2$,
5—$\eta_0 = 0.7$, $\zeta_0 = 0.01$, $\gamma = 5/3$

Instability can arise if the entropy density decreases outward. This fact can be used to explain experimentally observed turbulence in vapor plumes. Physically, the instability mechanism is similar to that responsible for convective instability in static stratified media when the entropy decreases in the upward direction.[6]

It is easy to see that the group-invariant solution of gas-dynamics equations (3.22) can be generalized to take into account a nonuniform entropy distribution over the initial cloud. One can confirm that solutions similar to (3.22) exist when the density and pressure profiles have the form:

$$\begin{aligned}\rho(x,y,z,t) &= \rho_0(t)\,[\psi(x,y,z,t)]^k \\ p(x,y,z,t) &= p_0(t)\,[\psi(x,y,z,t)]^{k+1}\end{aligned} \quad (3.28)$$

During the adiabatic expansion the value:

$$S = p/\rho^\gamma = (p_0/\rho_0^\gamma)\psi^{1-k(\gamma-1)} \quad (3.29)$$

is conserved in each Lagrangian fluid element. The adiabaticity condition $\partial S/\partial t + \mathbf{u}\cdot\nabla S = 0$ is satisfied if $p_0 = C\rho_0^\gamma$, where C is a constant. According to (3.29), the entropy gradient in the expanding vapor cloud is directed outward (inward) if $k(\gamma-1) > 1$ ($k(\gamma-1) < 1$). For a uniform initial distribution of entropy $k = 1/(\gamma-1)$.

Notice that the expansion is nonsteady; each fluid element is accelerated during the expansion. In a frame connected with a particular fluid element this acceleration is equivalent to a gravitational force — $\delta M d^2\mathbf{R}/dt^2$, δM

being the mass of the fluid element. It is known from hydrostatics[6] that the mechanical equilibrium of a fluid in a gravitational field is unstable if the entropy gradient is parallel to the gravity force. We can expect, therefore, that the expanding vapor plume will be unstable when $k < 1/(\gamma - 1)$, i.e., when the entropy of vapor in the external part of the plume is smaller than that in the central part of the plume. Destabilization takes place because of the buoyancy experienced by fluid elements in the nonuniform inertial force field.

Linear analysis of stability of a spherical cloud expanding into vacuum was performed by Book.[25] It was found that the initial perturbations grow for all spherical harmonics $l > 0$, provided $k < 1/(\gamma - 1)$. The relative size of the perturbations vary asymptotically like the unperturbed radius. At early times, however, when the acceleration is not small, the perturbations grow quickly. If $l \gg 1$ there is almost exponential growth for $t < R_0(M/E)^{1/2}$, the number of e-foldings during this time being of the order of $l^{1/2}$. The total amplification and the time required to approach the asymptotic state when the perturbations "freeze out" both increase with l. Note that as $l \to \infty$, the instability growth rate diverges. This means that the problem is not mathematically well posed. Physically, dissipative phenomena set an upper limit on the instability growth rate.

3.6 Dynamics of vapor expansion into an ambient gas

When the vapor plume produced by laser ablation expands into a gaseous medium, the ablated material acts on the surrounding gas as a piston. As a result, a shock wave is generated in front of the contact surface. During the expansion of the plume the vapor pressure drops and approaches the pressure of the ambient atmosphere. The shock wave degenerates into a sound wave, and the contact surface stops moving. This stage is known as the stationary plume. At this stage the pressure gradient inside the plume is small, and the pressure at the contact surface is equal to the ambient pressure, p_0. In this section we will study the effect of various parameters on the shape of the vapor plume. We will use the special solution of gas dynamics equations considered in Section 3.4. Strictly speaking, this solution is valid only for the expansion of gas into vacuum. However, the estimate of the plume shape on the basis of this solution yields a reasonable agreement with experiments.

We consider an adiabatic expansion of ellipsoidal vapor cloud with initial dimensions $2X_0$, $2Y_0$, and Z_0. We take the density and pressure profiles in the form (3.22). To estimate the shape of the stationary plume, we study the motion of the isobaric surface:

$$p(x, y, z, t) = p_0 \qquad (3.30)$$

CHAPTER 3. PROCESSES IN THE VAPOR PLUME

Using (3.22) we can rewrite (3.30) as

$$\tilde{p}_0 = (\xi\eta\zeta)^{-\gamma}\left(1 - \frac{\tilde{x}^2}{\xi^2} - \frac{\tilde{y}^2}{\eta^2} - \frac{\tilde{z}^2}{\zeta^2}\right)^{\gamma/(\gamma-1)} \qquad (3.31)$$

where $\tilde{p}_0 = p_0 I_2(X_0 Y_0 Z_0)/E$, $\tilde{x} = x/X_0$, $\tilde{y} = y/Y_0$, $\tilde{z} = z/Z_0$. It can be shown that the distance of each point on the surface (3.31) from the origin reaches its maximum at some instant t^*. We illustrate this in the simplest case—a spherical cloud. The radius $R = (\tilde{x}^2 + \tilde{y}^2 + \tilde{z}^2)^{1/2}$ of the isobaric surface satisfies the equation:

$$R^2(\tau) = \xi^2(\tau)\left[1 - \tilde{p}_0^{(\gamma-1)/\gamma}\xi(\tau)^{3(\gamma-1)}\right] \qquad (3.32)$$

The equation of motion for $\xi(\tau)$ follows from (3.26):

$$\xi' = \left[\frac{2}{3(\gamma-1)}\left(1 - \xi^{3(1-\gamma)}\right)\right]^{1/2}$$

$$\xi(0) = 1$$

The condition for the existence of a real solution of Equation (3.32) is $\tilde{p}_0 < 1$. Differentiating Equation (3.32), we find

$$2R\,dR/d\tau = \xi\xi'\left[2 - (3\gamma-1)\xi^{3(\gamma-1)}\tilde{p}_0^{(\gamma-1)/\gamma}\right]$$

The derivative $dR/d\tau$ is positive at $\tau = 0$, provided

$$\tilde{p}_0 < [2/(3\gamma-1)]^{\gamma/(\gamma-1)} < 1 \qquad (3.33)$$

If the inequality (3.33) is fulfilled, $R(\tau)$ reaches its maximum value at $\tau > 0$. The maximum radius of the isobaric surface is in this case

$$R_{max} = [3(\gamma-1)/(3\gamma-1)]^{1/2}[2/(3\gamma-1)]^{1/3(\gamma-1)}\tilde{p}_0^{-1/3\gamma}$$

Otherwise, if $[2/(3\gamma-1)]^{\gamma/(\gamma-1)} < \tilde{p}_0 < 1$, the maximum of $R(\tau)$ is reached at $\tau = 0$.

The general three-dimensional case can be investigated in a similar way. Equation (3.31) must in this case be solved numerically. The results of calculations and the comparison with experiment can be found in Stangl et al.[26] Although the model under discussion is extremely simplified, it correctly describes the dependence of the plume length on the gas pressure.

The above consideration shows that the contact surface that separates the ablated material from the gaseous ambient is decelerated during its motion and stops at $R = R_{max}$. In the coordinate frame that moves with the contact boundary, this deceleration is equivalent to a gravitational force directed away from the ablation products toward the shock-compressed gas. The pressures are the same on both sides of the contact surface, while the temperature of the shock-compressed ambient gas, under certain conditions, may be higher than that of the adiabatically expanding vapor; therefore, the vapor density may be higher than the gas density. In this case the contact boundary is unstable. This instability is similar to the Rayleigh-Taylor instability (RTI) of static equilibrium between two fluids of different densities in a gravitational field. The equilibrium is unstable when the density of the upper fluid is higher than that of the lower fluid. The RTI is encountered in widely different situations. The most well known examples are: supernova explosions, inertial confinement fusion, detonation, cavitation, cumulation, and magnetic field compression. The instability of the contact boundary between an expanding vapor plume and a surrounding gas is quite similar to the instability of the interface between the detonation products and the air in a spherical explosion. A linear analysis of this instability was performed by Anisimov and Zel'dovich.[27] Near the contact boundary, the scale length of the unperturbed flow is $h \propto c_s^2/|g|$, where c_s is the sound speed and $g = d^2R/dt^2$ is the acceleration of contact surface. Let us examine the evolution of fast-growing perturbations with wavelength $\lambda \ll h$. During the perturbation growth time, the acceleration g and the density of heavy (ρ_1) and light (ρ_2) fluids all remain practically constant, so that the instability growth rate is a slowly varying function of the time. Assuming the perturbations of hydrodynamic variables to be proportional to the factor $P_l(\cos\theta) \times \exp(\int \gamma dt)$, where P_l is the Legendre polynomial, and considering the limiting case $l \gg 1$, we obtain a set of linear equations that can be reduced to two equations for the velocity perturbation potential:

$$\varphi_i(r,\theta,t) = f_i(r) \times P_l(\cos\theta) \times \exp(\int \gamma\, dt)$$
$$f_i'' + F_i(r)f_i' + \left[\gamma^2/c_{si}^2 + l^2/R^2\right] f_i = 0 \qquad (3.34)$$

Here $F_i(r) = d\log(\rho_i)/dr$, $\rho_i(r)$ is the unperturbed density profile, $i = 1$ refers to the vapor, and $i = 2$ refers to the surrounding gas. The quantities ρ_i, c_{si}, R, and γ are slowly varying functions of time. The boundary conditions for Equations (3.34) follow from the continuity of the pressure and normal velocity component at the (perturbed) contact boundary. Solving Equations (3.34) in the semiclassical approximation and imposing the boundary conditions, we obtain the dispersion equation for short-wavelength perturbations in the form:

$$\gamma^2 = At\left\{l\,|\,g/R\,| + g^2(1 - At^2)(c_{s1}^{-2} - c_{s2}^{-2})/4\right\} \qquad (3.35)$$

where $At = (\rho_1 - \rho_2)/(\rho_1 + \rho_2)$ is the Atwood number. The first term in the parentheses in Equation (3.35) corresponds to the limiting case of an incompressible fluid; the second term takes into account the compressibility (to a first approximation in λ/h). In the limit $l \gg 1$, this term does not depend on l. Depending on the ratio of the acoustic velocities, the compressibility can have either a stabilizing or destabilizing effect on the contact boundary motion.

According to Equation (3.35), the instability growth rate diverges as $l \to \infty$. As in Section 3.5, the instability problem is not mathematically well posed. The dissipation and surface tension (which are not taken into account in our analysis) lead to the stabilization of short-wavelength perturbations.

The RTI of the contact surface is observed experimentally.[14,28] Estimates show that under typical experimental conditions, a substantial growth of perturbations with $l \propto 10\text{--}50$ can be expected. These short-wavelength perturbations do not lead to an important asymmetry of the vapor expansion. The primary manifestation of a short-wave instability is the mixing of the vapor with the ambient gas near the contact surface. The intermixed layer conserves the symmetry of the unperturbed flow. The mixing process can be described as turbulent diffusion. A detailed analysis of this problem is given in reference 29.

REFERENCES

1. Frenkel, J., *Kinetic Theory of Liquids*, Dover Publ., New York, 1955

2. Landau, L. D. and Lifshitz, E. M., *Statistical Physics, Part I*, Pergamon Press, Oxford, 1980

3. Anisimov, S. I., Vaporization of metal absorbing laser radiation, *Sov. Phys.-JETP*, **27**, 182, 1968

4. Tamm, I. E., Width of high-intensity shock waves, In: *Quantum Field Theory and Hydrodynamics, Proc. of P. N. Lebedev Physics Institute*, vol. **29**, Ed. by Skobel'tsyn, D. V., Consultants Bureau, New York, 1967, p. 231

5. Mott-Smith, H. M., The solution of the Boltzmann equation for a shock wave, *Phys. Rev.*, **82**, 885, 1951

6. Landau, L. D. and Lifshitz, E. M., *Fluid Mechanics*, Pergamon Press, Oxford, 1987

7. Kogan, M. N. and Abramov, A. A., Direct simulation solution of the strong evaporation and condensation problem, In: *Rarefied Gas Dynamics*, Ed. by Beylich A. E., VCH, Weinheim, 1991, p. 1251

8. Bishaev, A. M. and Rykov, V. A., Recondensation of a monatomic gas at low Knudsen numbers, *U.S.S.R. Comput. Math. and Mathematical Physics*, **18**, N 3, 182, 1978

9. Anisimov, S. I. and Rakhmatulina, A. Kh., The dynamics of the expansion of a vapor when evaporated into a vacuum, *Sov. Phys.-JETP*, **37**, 441, 1973

10. Murakami, M. and Oshima, K., Kinetic approach to the transient evaporation and condensation problem, In: *Rarefied Gas Dynamics*, Vol. **2**, Ed. by Becker, M. and Fiebig, M., DFVLR Press, Porz-Wahn, 1974, p. 6

11. Anisimov, S. I., Imas, Ya. A., Romanov, G. S., and Khodyko, Yu. V., *Action of High-Power Radiation on Metals*, National Technical Information Service, Springfield, Virginia, 1971

12. Zeldovich, Ya. B. and Raizer, Yu. P., *Physics of Shock Waves and High-Temperature Hydrodynamic Phenomena*, Academic Press, New York, 1966

13. Igoshin, V. I. and Kurochkin, V. I., Laser evaporation of metals in a gaseous atmosphere, *Sov. Journ. Quantum Electron.*, **14**, 1049, 1984

14. Scott, K., Huntley, J. M., Phillips, W. A., Clarke, J., and Field, J. E. Influence of oxygen pressure on laser ablation of $YBa_2Cu_3O_{7-x}$, *Appl. Phys. Lett.*, **57**, 922, 1990

15. Gilgenbach, R. M. and Ventzek, P. L. G., Dynamics of excimer laser-ablated aluminum neutral atom plume measured by dye laser resonance absorption photography, *Appl. Phys. Lett.*, **58**, 1597, 1991

16. Gupta, A., Braren, B., Casey, K. G., Hussey, B. W., and Kelly, R., Direct imaging of the fragments produced during excimer laser ablation of $YBa_2Cu_3O_{7-\delta}$, *Appl. Phys. Lett.*, **59**, 1302, 1991

17. Srinivasan, R., Ablation of polymethylmethacrylate films by pulsed (ns) ultraviolet and infrared (9.17 μm) lasers: a comparative study by ultrafast imaging, *Journ. Appl. Phys.*, **73**, 2743, 1993

18. Srinivasan, R., Ablation of polyimide (Kapton^{TM}) films by pulsed (ns) ultraviolet and infrared (9.17 μm) lasers. A comparative study, *Appl. Phys.*, **A56**, 417, 1993

19. Ovsiannikov, L.V., *Group Analysis of Differential Equations*, Academic Press, New York, 1982

20. Ovsiannikov, L.V., New solutions of hydrodynamic equations, *Dokl. AN SSSR* (Russian), **111**, 1, 1956

21. Gradsteyn, I. and Ryzhik, I., *Tables of Integrals, Series, and Products*, Academic Press, New York, 1980

22. Anisimov, S. I. and Lysikov, Yu. I., Expansion of gas cloud in vacuum, *Journ. Appl. Math. and Mech.*, **34**, 882, 1970

23. Anisimov, S. I., Baeuerle, D., and Luk'yanchuk, B. S., Gas dynamics and film profiles in pulsed-laser deposition of materials, *Phys. Rev.*, **B48**, 12076, 1993

24. Anisimov, S. I., Luches, A., and Luk'yanchuk, B. S., Film profiles in pulsed-laser deposition of materials (in press)

25. Book, D. L., Linear stability of self-similar flow: 4. Convective instability of a spherical cloud expanding into vacuum, *NRL Memorandum Report 3852, Naval Research Laboratory*, Washington, 1978

26. Stangl, E., Luk'yanchuk, B. S., Schieche, H., Piglmayer, K., Baeuerle, D., and Anisimov, S., Dynamics of the vapor plume in laser materials ablation, *NATO Advanced Study Institute, Excimer Lasers*, Crete, Greece, September, 1993

27. Anisimov, S. I. and Zel'dovich, Ya. B., Rayleigh-Taylor instability of the interface between the detonation products and a gas in a spherical explosion, *Sov. Tech. Phys. Lett.*, **3**, 445, 1977

28. Braren, B., Casey, K. G., and Kelly, R., On the gas dynamics of laser-pulse sputtering of polymethylmethacrylate, *Nucl. Instrum. & Methods*, **B58**, 463, 1991

29. Anisimov, S. I., Zel'dovich, Ya. B., Inogamov, N. I., and Ivanov, M. F., The Taylor instability of contact boundary between expanding detonation products and a surrounding gas, In: *Shock Waves, Explosions and Detonation, AIAA Progress in Astronautics and Aeronautics Series*, Vol. **87**, 218, 1983

Chapter 4

EFFECTS OF ULTRASHORT LASER PULSES

In this chapter we shall concentrate on the interaction of picosecond and sub-picosecond laser pulses with matter. The problems related to this interaction have attracted intense interest since the invention of picosecond lasers. The problem of electron relaxation dynamics is of particular interest due to important applications of ultrafast laser technology, such as the ultrafast desorption of surface adsorbates, generation of ultrashort X–ray pulses, ultrafast melting and annealing.

We will describe the electron and lattice temperature dynamics in a metal absorbing an ultrashort laser pulse, the laser-induced electron emission, and anomalous emission of short-wavelength radiation from metal surfaces. In this analysis, we will consider a metal to be a two-temperature system, with different electron and lattice temperatures. We will briefly discuss some of the recent experimental studies of hot-electron transport and relaxation in metals, which confirm the applicability of this simple model to the study of interaction of ultrashort laser pulses with metals.

4.1 Electron and lattice temperature dynamics in a metal heated by ultrashort laser pulses

At energy deposition by an ultrashort visible (or near IR) laser pulse, most of the energy is absorbed by "free" electrons through inverse Bremsstrahlung. The energy then is thermalized within the electron subsystem. The heating of the lattice takes a longer time due to the large ion-to-electron mass ratio. As first was indicated in references 1 and 2, the electron gas excited by an ultrashort laser pulse can be driven to transient nonequilibrium with the lattice. In this situation, the metal can be considered a two-temperature system and is described by the following set of equations:[1,2]

$$\begin{cases} c_e \partial T_e/\partial t = \partial/\partial z(\kappa \partial T_e/\partial z) - \alpha(T_e - T_i) + Q \\ c_i \partial T_i/\partial t = \alpha(T_e - T_i) \end{cases} \quad (4.1)$$

Here c_e and c_i are the heat capacities per unit volume of the electron and ion subsystems, respectively; T_e and T_i are the temperatures of electrons and the lattice; and α is the parameter characterizing the energy exchange rate between the electron and lattice subsystems. We neglect the phonon component of thermal conductivity, since it is much smaller in metals than is the electron component. The laser energy deposition Q is taken (as in Chapter 1) in the simplest form:

$$Q = I(t)(1 - R)\mu \exp(-\mu z) \quad (4.2)$$

We assume that the laser beam travels along the z-axis and that the surface of the metal is located at $z = 0$.

At electron temperatures $k_B T_e \ll E_F$, where E_F is the Fermi energy, the electron thermal conductivity κ in the nonequilibrium two-temperature metal can be approximated as follows:[3]

$$\kappa = \kappa_0 T_e/T_i \quad (4.3)$$

where κ_0 = constant is the thermal conductivity of metal in equilibrium ($T_e = T_i$). The electron heat capacity c_e depends on the electron temperature; at $k_B T_e \ll E_F$, $c_e \approx \beta T_e \ll c_i$, $\beta = \pi^2 n_e k_B^2/2E_F$. The energy exchange rate α was determined in references 4 and 5. It was shown that α is proportional to the electron-phonon coupling constant and can be estimated from the electrical conductivity measurements. Estimates for different metals give values of α on the order of 10^{10}–10^{12} W/cm^3 K.

Equations (4.1) contain two characteristic times, associated with the energy exchange between electrons and the lattice: $\tau_1 = c_e/\alpha$—the time of electron cooling, and $\tau_2 = c_i/\alpha$—the time of lattice heating. If the laser pulse length τ_l is shorter than τ_2, the lattice remains cold during the laser pulse. If $t \ll \tau_1$, the electron-lattice energy exchange term may be neglected in both equations (4.1). In the latter case, Equation (4.1) can be solved easily. Solving for the surface temperature, $T_e(0, t)$, gives:

$$T_e(0, t) = \left[T_0^2 + \frac{2A\mu}{\beta} \int_0^t I(t-s) \exp(ps) \operatorname{erfc}(ps)^{1/2} ds \right]^{1/2}$$

where $p = \mu^2 \kappa_0/\beta T_0$ and T_0 is the initial temperature. At later times, $t \gg \tau_1$, the term $c_e \partial T_e/\partial t$ in the first equation (4.1) can usually be neglected due to small electronic heat capacity. At the same time, one also can neglect the lattice temperature, which is much smaller than the electron temperature, $T_i \ll T_e$, if $t \ll \tau_2$. The set of equations (4.1) can thus be reduced to:

$$\partial/\partial z(\kappa \partial T_e/\partial z) - \alpha T_e + AI(t)\mu \exp(-\mu z) = 0$$
$$c_i \partial T_i/\partial t - \alpha T_e = 0 \qquad (4.4)$$

At low laser intensities, when the parameter $\kappa_0 \mu^3 AI/\alpha^2 T_0 \ll 1$, the cooling of electrons is mainly due to energy exchange with the lattice. We can ignore the heat conduction term in Equation (4.4), and obtain the following simple solution for the electron temperature:

$$T_e(z,t) = I(t)(A\mu/\alpha)\exp(-\mu z) \qquad (4.5)$$

In the opposite case, $\kappa_0 \mu^3 AI/\alpha^2 T_0 \gg 1$, the cooling of electrons and the spatial profile of electron temperature are determined by electron thermal conduction. The solution of Equation (4.4) may be written in the form:

$$T_e(z,t) = T_e(0,t)(1 - z/z_0)^2 \qquad (4.6)$$

where $T_e(0,t) = [3T_0 A^2 I^2(t)/2\alpha\kappa_0]^{1/3}$, and $z_0 = [18\kappa_0 AI(t)/\alpha^2 T_0]^{1/3}$. When $t \gg \tau_2$, we have $T_e \approx T_i$, and (4.1) is reduced to the one-temperature heat-conduction problem considered in Chapter 1. Approximate analytical solutions of Equations (4.1) in various limiting cases were obtained and analyzed in references 6–9.

Generally, Equations (4.1) can be solved numerically. There are several works in which such numerical calculations have been performed to interpret experimental results.[10-12] It was found that the model under consideration provides a good quantitative description of hot-electron dynamics in metals.

4.2 Anomalous electron and light emission from metals heated by ultrashort laser pulses

Laser-pulse excitation of a metal surface can lead to the emission of electrons through two mechanisms: (1) the multiphoton photoelectric effect and (2) thermionic emission. Both emission mechanisms have been studied extensively in the nanosecond pulse length range,[13] when the lattice temperature T_i was in equilibrium with the electron temperature T_e. These experiments have demonstrated that the multiphoton photoemission dominates thermionic emission for incident fluences less than ~ 0.1 J/cm^2. Laser fluences above this level lead to a significant contribution from thermionic emission.[14,15] In experiments with picosecond laser pulses, multiphoton photoemission has been observed[16] primarily at laser fluences below the onset of surface damage. The situation changes when excitation times cross into the subpicosecond range. The electron subsystem becomes briefly uncoupled from the lattice, enabling T_e to become larger than T_i.[1,2,10] Because the heat capacity of electrons is substantially less than that of the lattice, the degree of nonequilibrium electron heating can be very high. According to Anisimov et al.,[2] the dividing line between multiphoton and thermionic emission is estimated to be about 1 mJ/cm^2 for 100 fs pulses. This estimate is supported by experiments.[10,17]

Multiphoton electron emission is an instantaneous process, i.e., the current pulse shows no delay relative to the laser pulse. For thermionic emission, the shape of a current pulse is governed by the time dependence of the target temperature. In the case of nanosecond pulses, the maximum of the current pulse is delayed relative to the laser pulse. Its duration exceeds the laser pulse length τ_l and the time of delay depends on the thermal properties of the target material. With subpicosecond pulses, the electron temperature follows almost instantaneously after the laser radiation intensity. The delay of thermionic emission pulse relative to the laser pulse is on the order of $\tau_1 = c_e/\alpha$, which can be smaller than the laser pulse length. We therefore see that with ultrashort laser pulses, a combination of the two fundamental emission processes can be observed, whereupon electrons in the high-energy tail of the transient thermal distribution further absorb laser quanta and escape from the solid. This mechanism, called thermally assisted photoemission, was first considered by Anisimov et al.[13]

We now will discuss the effect of transient electron heating on the characteristics of emission current. If the laser intensity is not very high and the parameter $\Delta = e^2 |E_n|^2 / m\hbar\omega^3 \ll 1$, the perturbation theory can be used for the calculation of n-photon photoemission current density. The transition probability, as a function of the electron quasi-momentum, has been calculated by Kantorovich.[18] This probability should be averaged over the Fermi distribution of electrons. If the effects associated with heating are ignored, the calculation results in the dependence of photoemission current on the laser intensity in the form: $j_n \sim \Delta^n \sim I^n$, where n is the integer part of $(\varphi/\hbar\omega + 1)$, and φ is the work function, i.e., n is the minimum number of photons required to emit an electron at $T_e = 0$. The transient heating produces electrons with energies above the Fermi level, so emission can now take place as the result of absorption of fewer than n photons. The calculation of total emission current in this case was performed by Kantorovich[18] and Ansimov et al.[19] The partial currents considered as functions of the electron quasi-momentum and representing the absorption of n, $n - 1$, etc. photons were calculated numerically. Next, the results were averaged over a Fermi distribution with the temperature, which depends on laser intensity. As an example of such calculations, in Figure 4.1 the emission current from silver irradiated by a neodymium glass laser is shown as a function of laser intensity with the electron heating taken into account (solid curve). The dashed curve was obtained by ignoring this heating ($T_e = 0$). This line represents the dependence $j \propto I^5$. Real dependence exhibits a higher degree of nonlinearity, $n_{eff} = d\log j/d\log I > 5$.

In Figure 4.2 the dependence of the emission current on the laser frequency at different intensities is shown. The calculation was performed for a silver target.[18] We see that at low intensities, the dependence of the emission current on the light frequency is not monotonic. However, at high intensities, the electron heating mixes currents of different orders and this smooths out the

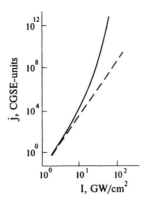

Figure 4.1: Photoelectric emission from silver. The solid curve was obtained when including the electron heating; the dashed curve, omitting this heating.

frequency dependence of the total current. Note that the degree of nonlinearity of the total current, $n_{eff} = d\log j/d\log I$, is also a non-monotonic function of laser intensity. It can be shown that $n_{eff}(\omega)$ has a number of maxima at "resonant" frequencies $\omega_k \approx \varphi/\hbar k$, where k is an integer. When the laser frequency is close to ω_k, the main contribution to the total emission current is due to k- and $k-1$-photon partial currents. The total current can be represented as:

$$j \approx B_k I^k + B_{k-1} I^{k-1} \exp(-\Omega/k_B T_e), \quad \Omega \approx \varphi - \hbar\omega(k-1)$$
$$k\hbar\omega > \varphi > (k-1)\hbar\omega$$

Using, for simplicity, $T_e = AI\mu/\alpha$ (see (4.5)), we obtain:

$$n_{eff} = k + \delta n, \qquad \delta n = (x-1)/(be^x + 1)$$

with $x = \alpha\Omega/AI\mu k_B$ and $b = B_k I/B_{k-1} \approx \Delta \ll 1$. It is easy to see that δn as a function of ω has a maximum. The maximum value of δn is $\delta n_{max} = x_m - 2$, where x_m satisfies the equation $(x_m - 2)b\exp(x_m) = 1$. For typical experimental conditions δn_{max} is on the order of several units. Anomalous behavior of the emission-current nonlinearity near the "resonant" frequency ω_k was observed experimentally by Lompre et al.[20] A bandwidth-limited 15 ps laser pulse was used to induce multiphoton photoelectric emission from a polycrystalline gold target. The transition from a four-photon to a five-photon emission was studied. A sharp increase in n_{eff} was observed, with maximum value $n_{eff} = 14$. The observed dependence of the degree of nonlinearity on the laser frequency is qualitatively similar to that predicted by Anisimov et al.[13] and Kantorovich.[21] However, in the experiments the change of n_{eff} takes place in a much narrower laser frequency range than was calculated by Kantorovich.[21] This difference is not yet well understood.

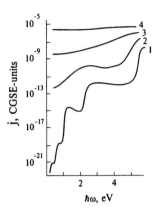

Figure 4.2: Dependence of the emission current on the photon energy for different laser intensities (W/cm^2): 1—10^9, 2—4×10^9, 3—7×10^{10}, 4—10^{11}. Surface reflectivity $R = 0.99$.

The estimates[2,11] and experimental measurements[12,22] show that the electron temperature in metals irradiated by ultrashort laser pulses can reach values about 1 eV, whereas the lattice temperature can remain lower than melting point. Under this condition, along with the phenomena of thermoemission and photoemission of electrons discussed above, thermal luminescence of the electron gas in the metal also may take place. A significant portion of the radiation will be in the visible and UV regions of the spectrum. The delay of the luminescence pulse relative to the laser pulse is expected to be on the order of τ_1.

Experimental observations of the luminescence of metals heated by ultrashort pulses was reported in references 8, 23, and 24. Temporal and spectral characteristics of the radiation emitted from the silver and tungsten surfaces, subjected to 10 and 40 ps laser pulses, were determined. The choice of metals used was dictated by the fact that the value of α for W is about two orders of magnitude higher than for Ag. Estimates show that in the experiments under discussion, electrons in W may be in thermal equilibrium with the lattice, while in Ag a substantial overheating of electrons with respect to the lattice can be expected. The measurements made at low laser fluences in the 560–720 nm spectral range show that the luminescence pulse duration in Ag is comparable to that of the laser pulse, while in W the luminescence pulse is several times longer.

The dynamics of the luminescence spectrum were studied at higher laser fluences for the silver target in the 510–560 nm spectral range. The most intense lines occurring at 521 and 546 nm belong to this range. The fluence was 1.0 J/cm^2, and the laser pulse length was 10 ps. Under these conditions, a noticeable evaporation of the target may be observed. The measurements show that during the initial 400 ps, the emission spectrum is continuous.

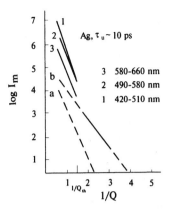

Figure 4.3: The luminescence intensities in various spectral bands versus the inverse laser fluence (solid lines). The broken lines a and b correspond to the $1/Q$-dependence of the black-body radiation at $T = T_e$ and $T = T_i$, respectively. The laser pulse duration is 10 ps, the time resolution is 6 ps.

Later, at $t \sim 0.5$–1 ns, bands appear, and at $t > 2$ ns the spectral lines of atomic silver are observed. Note that experiments of this type can provide important information on the kinetics of vaporization of solids and liquids.

Figure 4.3 shows the radiation intensity in three different spectral bands as a function of the laser fluence. This dependence is presented in the variables ($\log I_m, 1/Q$), where I_m is the maximum of radiation intensity and Q is the laser fluence (J/cm^2). The laser pulse duration is 10 ps; the time resolution is 6 ps. According to calculations[11] under the conditions of the experiment,[24] the maximum electron temperature is a linear function of the laser fluence absorbed. This means that the radiation intensity depends exponentially on $1/T_e$, which corresponds to the black- (grey-) body radiation. The broken lines in Figure 4.3 correspond to black-body radiation intensity at temperatures T_e (line a) and T_i (line b). We see, then, that at low laser fluences ($Q < Q_{th} \approx 0.6$ J/cm^2), the luminescence observed corresponds to the equilibrium emission of radiation by hot electrons. The emission signal is in this case "inertialess"; i.e., it follows the laser pulse shape. However, above the threshold fluence Q_{th}, the radiation intensity observed is higher than the black-body radiation intensity at temperature T_e. Several mechanisms for this phenomenon have been suggested in the literature. One of them is a breakdown of the equilibrium Fermi distribution in metals at high intensities. We shall discuss this mechanism briefly.

The distribution function of electrons in metals is described by the Boltzmann kinetic equation. The Fermi distribution function is the stationary solution of this equation, i.e., it causes vanishing of the collision integral. Kats et al.[25] showed that for the particles interacting in accordance with the power law $V(r) \propto r^{-\alpha}$ (Coulomb forces, Van der Waals forces, etc.), the collision integral vanishes also for power law distributions $f(\mathbf{p}) \propto p^s$. Similar distributions (spectra) are encountered, for example, in the theory of weak turbulence of plasma waves. This class of solutions of the kinetic equation is valid in a finite region of the momentum space $p_1 < p < p_2$, and presupposes the existence of an energy source and an energy sink outside this region. The properties of systems with such distribution functions, especially those properties that are sensitive to the presence of high-energy "tail" particles, differ greatly from those of equilibrium systems. The analysis carried out by Kats et al.[25] shows that in the case of Coulomb interaction, the distribution function of the form $f(\mathbf{p}) \propto p^{-5/2}$ corresponds to constant energy flux in the momentum space and is local, i.e., the collision integral converges at both small and large momenta.

In plasma wave turbulence, the power law spectra are generated when the pumping intensity exceeds a certain critical level. Above this level, the spectrum is represented by the superposition of the equilibrium (e.g., Rayleigh–Jeans) spectrum and the power law spectrum. Similarly, in the case of electrons in a strong radiation field, the power law electron distributions can be expected to appear at sufficiently high laser intensities. A detailed analysis of experimental data is necessary to prove (or disprove) this mechanism of anomalous emission of radiation from metals heated by ultrashort laser pulses.

REFERENCES

1. Anisimov, S. I., Imas, Ya. A., Romanov, G. S., and Khodyko, Yu. V., *Action of High-Power Radiation on Metals*, National Technical Information Service, Springfield, VA, 1971

2. Anisimov, S. I., Kapeliovich, B. L., and Perel'man, T. L., Electron emission from metal surfaces exposed to ultrashort laser pulses, *Sov. Phys.–JETP*, **39**, 375, 1974

3. Agranat, M. B., Benditskii, A. A., Gandel'man, G. M., Devyatkov, A. G., Kondratenko, P. S., Makshantsev, B. I., Rukman, G. I., and Stepanov, B. M., Noninertial radiation from metals in interaction with ultrashort pulses of coherent infrared radiation, *JETP Lett.*, **30**, 167, 1979

4. Kaganov, M. I., Lifshitz. I. M., and Tanatarov, L. V., Relaxation between electrons and the crystalline lattice, *Sov. Phys.–JETP*, **4**, 173, 1957

5. Allen, P. B., Theory of thermal relaxation of electrons in metals, *Phys. Rev. Lett.*, **59**, 1460, 1987

6. Brorson, S. D., Kaseroonian, A., Moodera, J. S., Face, D. W., Cheng, T. K., Ippen, E. P., Dresselhaus, M. S., and Dresselhaus, G., Femtosecond room-temperature measurements of the electron-phonon coupling constant λ in metallic superconductors, *Phys. Rev. Lett.*, **64**, 2172, 1990

7. Corkum, P. B., Brunel, F., Sherman, N. K., and Srinivasan-Rao, T., Thermal response of metals to ultrashort-pulse laser excitation, *Phys. Rev. Lett.*, **61**, 2886, 1988

8. Agranat, M. B., Anisimov, S. I., Ashitkov, S. I., Makshantsev, B. I., and Ovchinnikova, I. B., Thermal radiation emitted by metals due to disturbance of an equilibrium between electrons and the lattice, *Sov. Phys.-Solid State*, **29**, 1875, 1987

9. Anisimov, S. I. and Barsukov, A. V., Nonequilibrium heating of a metal by picosecond laser pulses, *Sov. Tech. Phys. Lett*, **17**, 607, 1991

10. Fujimoto, J. G., Liu, J. M., Ippen, E. P., and Bloembergen, N., Femtosecond laser interaction with metallic tungsten and nonequilibrium electron and lattice temperatures, *Phys. Rev. Lett.*, **53**, 1837, 1984

11. Anisimov, S. I., Makshantsev, B. I., and Barsukov, A. V., Metal surface heating by picosecond laser pulses, *Opt. and Acoust. Rev.*, **1**, 251, 1991

12. Wang, X. Y., Riffe, D. M., Lee, Y.-S., and Downer, M. C., Cooling dynamics of highly excited electrons in gold, *Phys. Rev. Lett.* (to be published)

13. Anisimov, S. I., Benderskii, V. A., and Farkas, G., Nonlinear photoelectric emission from metals induced by a laser radiation, *Sov. Phys.-Uspekhi*, **20**, 467, 1977

14. Farkas, Gy., Kertèsz, I., Nàrai, Zs., and Varga, P., On the intensity dependence of the non-linear electron emission from silver induced by a high power laser beam, *Phys. Lett.*, **A24**, 475, 1967

15. Farkas, Gy., Nàrai, Zs., and Varga, P., Dependence of non-classical electron emission from metals on the direction of polarization of laser beams, *Phys. Lett.*, **A24**, 134, 1967

16. Farkas, Gy., Horvàth, Z. Gy., and Kertèsz, I., Influence of optical emission on the nonlinear photoelectric effect induced by ultrashort laser pulses, *Phys. Lett.*, **A39**, 231, 1972

17. Schoenlein, R. W., Fujimoto, J. G., Easley, G. L., and Capehart, T. W., Femtosecond studies of image-potential dynamics in metals, *Phys. Rev. Lett.*, **61**, 2596, 1988

18. Kantorovich, I. I., Nonlinear surface photoelectric effect in metals subjected to intense light, *Sov. Phys.-Tech. Phys.*, **22**, 397, 1977

19. Anisimov, S. I., Inogamov, N. A., and Petrov, Yu. V., Intensity dependence of laser-induced electron emission current from metal surface, *Phys. Lett.*, **A55**, 449, 1976

20. Lompre, L.-A., Mainfray, G., Manus, C., Thibault, J., Farkas, Gy., and Horvàth, Z. Gy., A new effect in multiphoton photoeffect of a gold surface induced by picosecond laser pulse, *Appl. Phys. Lett.*, **33**, 124, 1978

21. Kantorovich, I. I., The effect of heating of electrons by means of optical radiation on the nonlinear surface photo-effect in metals, *Sov. Tech. Phys. Lett.*, **3**, 230, 1977

22. Riffe, D. M., Wang, X. Y., Downer, M. C., Fisher, D. L., Tajima, T., Erskine, J. L., and More, R. M., Femtosecond thermionic emission from metals in the space-charge-limited regime, *Journ. Opt. Soc. Amer.*, **B10**, 1424, 1993

23. Agranat, M. B., Anisimov, S. I., and Makshantsev, B. I., Anomalous thermal radiation from metals subjected to picosecond laser pulses, *Sov. Phys.-Solid State*, **29**, 1966, 1987

24. Agranat, M. B., Anisimov, S. I., and Makshantsev, B. I., The anomalous thermal radiation from metals produced by ultrashort laser pulses, *Appl. Phys.*, **B47**, 209, 1988; **B55**, 451, 1992

25. Kats, A. V., Kontorovich, V. M., Moiseev, S. S., and Novikov, V. E., Power law solutions of the Boltzmann kinetic equation, describing the spectral distribution of particles with fluxes, *JETP Lett.*, **21**, 5, 1975

Chapter 5

INSTABILITY OF SUBLIMATION WAVES IN SOLIDS: LINEAR THEORY

5.1 Corrugation instability of stationary evaporation waves in highly absorbing materials

In this section we will study the stability of planar sublimation waves in strongly absorbing solids. We will show that under certain conditions a planar, steady-state vaporization front appears to be unstable with respect to small perturbations, whose wave vector lies in the front plane. Nonlinear development of this instability leads to the formation of nontrivial evaporation regimes.

Although it is more common to investigate evaporation experimentally from a liquid phase, we will begin with the study of the solid-to-vapor phase transition for several reasons. First, the mechanism leading to the instability in the case of solid-to-vapor phase change works also in the case of liquid-to-vapor phase change; however, the analysis in the latter case is more complicated. Second, in materials with a rather high saturated vapor pressure at the melting point, the layer of liquid near the vaporization front is very thin and has only a slight effect on the development of instability. Third, a number of materials—e.g., carbon, several plastics and ceramics—do not form a liquid phase under conditions typical for laser experiments. A detailed linear analysis of instability of the liquid–vapor phase boundary will be given in Chapter 7.

The instability under consideration plays an important role at moderate laser intensities, $I \approx 10^5$–10^8 W/cm^2. As was mentioned in Chapter 1, a sharp boundary exists between the vapor and condensed phases at these intensities.

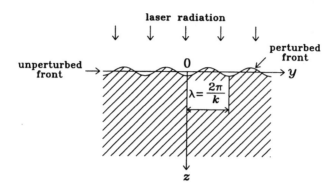

Figure 5.1: Schematic showing the vaporization front perturbations.

The velocity of this boundary, as well as the temperature distribution in the condensed phase were calculated in Chapter 1 for the stationary regime of evaporation. Now we shall examine the stability of this boundary with respect to small perturbations.[1] The situation is shown schematically in Figure 5.1. We will introduce small periodic perturbations to the vaporization front position and to the temperature field, and will study their evolution over time. As in Chapter 1, we will describe the temperature field by the heat conduction equation

$$c\rho \partial T/\partial t = \kappa \Delta T + Q \qquad (5.1)$$

where $Q = A\mu I \exp[\mu(Z-z)]$, and $Z = Z(x,y,t)$ is the position of the solid-gas phase boundary. We use below the same notations as in Chapter 1. Boundary conditions for Equation (5.1) can be written as follows. As $z \to \infty$, the temperature T approaches T_0 (we will assume here $T_0 = 0$). On the evaporation front $z = Z(x,y,t)$, we must impose the continuity condition for the normal component of the energy flux. This condition reads:

$$\kappa \nabla T = \rho V_n \Delta H \qquad (5.2)$$

where V_n is the normal component of evaporation front velocity, and ΔH is the difference between specific enthalpies of gaseous and condensed phases. As in Chapter 1, we can set the enthalpy jump equal to the specific heat of vaporization, $\Delta H \simeq L_{eff}$. However, for the nonplanar phase boundary, the value of L_{eff} depends on the local curvature of a phase boundary, and can be represented as follows:

$$L_{eff} = L_v - (\sigma/\rho)(1/R_1 + 1/R_2) \qquad (5.3)$$

Here, L_v is the specific heat of vaporization at the plane boundary, σ is the surface energy density (surface tension) for the solid-gas boundary, R_1 and R_2 are the principal radii of curvature of the boundary, and the normal is assumed to be directed toward the condensed phase. As we will see later,

CHAPTER 5. INSTABILITY OF SUBLIMATION WAVES

the dependence of the specific heat of vaporization on the phase boundary curvature results in the stabilization of short-wavelength perturbations.

The evaporation front velocity in Equation (5.2) is determined by the kinetics of the transition of atoms from the condensed phase to the gas. When a solid evaporates into vacuum, the mass flux is proportional to the saturated vapor pressure, and we can use Equation (1.16) for V_n.

We begin our analysis by studying the stability of the stationary evaporation wave. The temperature profile in this wave is described by Equation (1.19). The temperature and the stationary velocity V_s of the solid–gas boundary can be obtained from the solution of Equations (1.20) and (1.21). Introducing new variables:

$$\xi = \mu x \qquad \eta = \mu y \qquad \zeta = \mu(z - V_s t) \qquad \tau = tV_s^2/\chi \qquad \beta = V_s/\mu\chi$$

$$\Theta(\xi,\eta,\zeta,\tau) = \kappa\mu T(x,y,z,t)/AI \qquad \delta Z(\xi,\eta,\tau) = \mu[Z(x,y,t) - V_s t]$$

we transform Equation (5.1) and boundary condition (5.2) into the following form:

$$\beta^2 \partial\Theta/\partial\tau = \beta\partial\Theta/\partial\zeta + \Delta\Theta + \exp(\delta Z - \zeta) \qquad (5.4)$$

$$\nabla\Theta = \rho V_n L_{eff}/AI \qquad (5.5)$$

According to (1.16), the normal component of the evaporation front velocity can be written as follows:

$$V_n = V_0 \exp\{-U_{eff}/\Theta[\xi,\eta,\delta Z(\xi,\eta,\tau),\tau]\} \qquad (5.6)$$

where $U_{eff} = ML_{eff}/k_B$. The planar stationary evaporation wave is described by the solution $\Theta_s(\zeta)$ of Equations (5.4) through (5.6), which depends on the variable ζ only. Performing simple calculations, we arrive at the result

$$\Theta_s(\zeta) = B_1 \exp(-\zeta) + B_2 \exp(-\beta\zeta) \qquad \delta Z = 0$$
$$B_1 = (\beta - 1)^{-1} \qquad B_2 = -\lambda - B_1/\beta \qquad \lambda = \rho\chi\mu L_0/AI \qquad (5.7)$$

which coincides with (1.19). As mentioned in Chapter 1, the stationary temperature defined by (1.19) reaches its maximum at finite distance from the phase boundary $\zeta = \zeta_m > 0$, i.e., the temperature gradient near the vaporization front is positive. Performing simple calculations, we obtain for the coordinate of the temperature maximum:

$$\zeta_m = (\beta - 1)^{-1} \log[1 + \lambda\beta(\beta - 1)]$$

The maximum temperature is given by:

$$\Theta_{sm} = \beta^{-1}[1 + \lambda\beta(\beta - 1)]^{1/(1-\beta)}$$

It is easy to see that under typical conditions the distance of the maximum from the surface is on the order of skin depth, μ^{-1}.

We will now investigate the stability of the stationary solution (5.7). We introduce small perturbations to the temperature and to the phase boundary position in the form:

$$\begin{aligned}\Theta(\xi,\eta,\zeta,\tau) &= \Theta_s(\zeta) + \delta\Theta & \delta\Theta &= f(\zeta)\exp(ik\eta + \gamma\tau) \\ \delta Z(\xi,\eta,\tau) &= \Xi\exp(ik\eta + \gamma\tau) & & \end{aligned} \quad (5.8)$$

Substituting (5.8) into (5.4) through (5.6) and leaving the terms linear in perturbations, we arrive at an eigenvalue problem for dimensionless perturbation growth rate, γ. If the real part of γ is positive, then the solution (5.7) appears to be unstable. One should note that the linear growth rate γ depends only upon the modulus of the wave vector, k. For this reason the dependence of $\delta\Theta$ and δZ on the variable ξ was omitted in (5.7).

To derive the linear equations for small perturbations, we first should project the boundary conditions (5.5) and (5.6) imposed on the perturbed evaporation front $\zeta = \delta Z(\eta,\tau)$ onto the plane $\zeta = 0$. For this purpose we expand the boundary conditions in the power series in terms of the small quantity δZ, and confine ourselves to linear terms. Performing simple calculations we obtain the following result:

$$f'' + \beta f' - (\beta^2\gamma + k^2)f = -\Xi\exp(-\zeta) \quad (5.9)$$

$$\left.\begin{aligned} f'(0) &= \Xi(\gamma\beta^2 + \beta\lambda\Lambda k^2 - \Theta_0''), \\ f(0) &= \Xi(\Theta_0^2\beta\gamma p^{-1} - \Theta_0' + \Theta_0\Lambda k^2) \end{aligned}\right\} \quad \text{as } \zeta = 0 \quad (5.10)$$

where $\Lambda = \sigma\mu/\rho L_0$, $\Theta_0 = \Theta_s(0)$, $\Theta_0' = \Theta_s'(0)$, etc. For a rough estimate of the order of magnitude of Λ, note that if we substitute in Equation (5.3) $R \simeq k^{-1} \simeq d$, d being the interatomic distance, then both terms on the right side on this equation must become on the same order. We immediately obtain the estimate $\Lambda \simeq \mu d \ll 1$.

Now, we can solve Equation (5.9) with the boundary conditions (5.10) to obtain a dispersion equation which defines the dependence $\gamma(k)$.[1] This equation can be written conveniently in parametric form as:

$$\begin{aligned}\gamma &= \frac{\alpha-\beta}{\beta}\frac{\Theta_0' - (\alpha-\beta+1)^{-1} - \alpha\Lambda(\alpha\Theta_0 - \Theta_0')}{\alpha\Theta_0(\Theta_0 p^{-1} - \Lambda\beta) + (\lambda\beta(1-\lambda\beta)}\\ k^2 &= \alpha(\alpha-\beta) - \beta^2\gamma \qquad \text{Re}\,\alpha > 0\end{aligned} \quad (5.11)$$

where α is a parameter and $p = \lambda cM/k_B$.

We begin our analysis of the dispersion equation (5.11) with the case of real values of α, which correspond to real values of γ. Consider, first, the denominator of Equation (5.11). Estimate the magnitude of the product $\Lambda\beta$. Using the approximation $\Lambda \sim \mu d$ and changing to dimensional variables, we find that $\Lambda\beta \sim V_s d/\chi$. But χ/V_s is the spatial scale of the temperature field in the stationary evaporation wave (see (5.7)). For a macroscopic description of the problem to make sense, this quantity should be much greater than the interatomic distance. We see, hence, that $\Lambda\beta \ll 1$, and $1 - \Lambda\beta > 0$. One can show, also, that the product $\Lambda\beta \ll \Theta_0 p^{-1}$. Indeed, the inequality $\Theta_0 \gg p\lambda\beta$ can be rewritten as

$$\Theta_0 p^{-1} \exp(p\Theta_0^{-1}) \gg \lambda V_0/\mu\chi \approx V_0 d/\chi \qquad (5.12)$$

A typical value of d is $\simeq 10^{-8}$ cm, $c_0 \simeq 10^5$ cm/s, and χ is not less than 10^{-2} cm^2/s for condensed material. We see that the right side of Equation (5.12) is small compared with unity, which makes the inequality obvious. Thus, we see that the denominator in Equation (5.11) is positive.

We will now investigate the numerator of Equation (5.11). First, we see that, when $\alpha = \beta$, $\gamma = k = 0$. Further, since $\Theta_0' = \lambda\beta < 1$, we have at $0 < (\alpha - \beta) \ll 1$

$$\Theta_0' - (\alpha - \beta + 1)^{-1} \approx \lambda\beta - 1 < 0 \qquad (5.13)$$

It follows that $\gamma(k) < 0$ for small k, i.e., the long wavelength perturbations decay with time. As α increases, the magnitude of the term $(\alpha - \beta + 1)^{-1}$ decreases and when

$$\alpha = \alpha_1 = \beta - 1 + 1/\Theta_0' \qquad (5.14)$$

the left side of Equation (5.13) vanishes. If the latter term in the numerator $\alpha\Lambda(\alpha\Theta_0 - \Theta_0')$ is small at $\alpha = \alpha_1$, in comparison with Θ_0', then the growth rate γ changes sign at $\alpha \simeq \alpha_1$. Positive value of $\gamma(k)$ means that a plane stationary evaporation wave becomes unstable.

At large α, the term $\alpha\Lambda(\alpha\Theta_0 - \Theta_0')$ is large in absolute value. Therefore, the growth rate γ again becomes negative, if $\alpha > \alpha_2$. A reasonable estimate for the magnitude of α_2 is:

$$\alpha_2 \simeq (\Theta_0'/\Lambda\Theta_0)^{1/2} \qquad (5.15)$$

If Λ is not too small and $\alpha_1\Lambda(\alpha_1\Theta_0 - \Theta_0') \geq \Theta_0'$, the stabilizing effect of the last term in the numerator of (5.11) asserts itself at $\alpha \simeq \alpha_1$, and $\gamma(k)$ may remain negative for all finite values of k.

Using the estimate for the magnitude of Λ, we may readily verify that $\Lambda\alpha_1$ is always small compared with unity. Hence, we may conclude that the plane stationary evaporation wave is stable when $\alpha_1^2\Lambda\Theta_0 > \Theta_0'$, and unstable when $\alpha_1^2\Lambda\Theta_0 \ll \Theta_0'$. In the latter case, small perturbations of plane vaporization front grow with time if their wave numbers lie in some finite interval $k_1 < k < k_2$, where $k_1 \simeq \alpha_1(\alpha_1 - \beta)$, $k_2 \simeq \alpha_2(\alpha_2 - \beta)$.

Returning to the solution of the stationary problem, one can show easily that Θ_0 decreases monotonically as the laser intensity increases. This means that the stabilizing term $\alpha^2\Lambda\Theta_0$ related to the surface energy also decreases with laser intensity increasing. Therefore, a threshold in laser intensity must exist, above which a plane stationary evaporation wave becomes unstable. The stability of this wave at laser intensities below the threshold is due entirely to the contribution of the surface energy to the latent heat of vaporization. If this contribution is neglected, short wavelength perturbations, which make the plane stationary evaporation wave unstable, always exist.

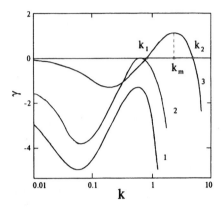

Figure 5.2: The dependence of the instability growth rate γ on the perturbation wave number k at different laser radiation intensities:
1—$I = 0.35 I^*$; 2—$I = I^*$; 3—$I = 3.33 I^*$; $I^* = 3.33 \times 10^{-3} \rho V_0 L_v$, $\Lambda = 10^{-3}$.

At laser intensities well above the instability threshold, upper and lower boundaries of the instability interval in k may be determined by Equations (5.14) and (5.15). Within this interval the dispersion equation (5.11) may be simplified as:

$$\gamma(k) = (p/\beta\Theta_0^2)[\Theta_0' - (k+1)^{-1} - k^2 \Lambda \Theta_0]$$

From the last equation, one can see that a maximum of the instability growth rate is reached at $k = k_m$, where:

$$k_m \simeq (2\Lambda\Theta_0)^{-1/3} \qquad (5.16)$$

and is equal to

$$\gamma_m = (p/\beta\Theta_0^2)\left[\Theta_0' - 1.9(\Lambda\Theta_0)^{1/3}\right] \qquad (5.17)$$

Detailed numerical investigations of the dispersion equation (5.11), made without simplifying assumptions, confirms the simple analysis described above. In Figure 5.2 the dispersion curves $\gamma(k)$ are shown corresponding to three different values of the laser intensity: subcritical $I < I^*$ (curve 1), critical $I = I^*$ (curve 2), and supercritical $I > I^*$ (curve 3). In the last case, instability growth rate is positive in a finite interval of wave numbers $k_1 < k < k_2$.

We now turn to Equation (5.17) to find the maximum growth rate γ_m. Because the negative term in (5.17), proportional to $\Lambda^{1/3}$, is significantly less than the positive term in the region of instability, we have the following estimate:

$$\gamma_m \approx p\Theta_0'/\beta\Theta_0^2 \qquad (5.18)$$

Transforming (5.18) to dimensional variables and using Equation (5.6), we obtain:

$$\tilde{\gamma}_m \approx (dV_s/dT_s)T_s' = dV_s/dz \qquad (5.19)$$

CHAPTER 5. INSTABILITY OF SUBLIMATION WAVES

We see from (5.19) that the instability growth rate is positive if a forward displacement of the evaporation front causes its velocity to increase.

We have studied the branch of dispersion equation (5.11), which corresponds to real γ and real k. This branch describes a nonoscillatory (aperiodic) growth of small initial perturbations. One can show, using Equation (5.11), that complex branches of the dispersion equation lie in a nonphysical domain for the case of strongly absorbing materials. For complex γ, the condition that the temperature perturbations must decrease at infinity, $\mathrm{Re}\,\alpha > 0$, is not satisfied. This means that only the aperiodic growth of small perturbations is possible in the case of sublimation of strongly absorbing solids. In Chapter 4 and in Section 5.3 of this chapter, we shall show that an oscillatory growth of small perturbations is possible for laser evaporation of liquids and laser sublimation of dielectrics.

5.2 Corrugation instability of nonstationary evaporation waves in highly absorbing materials

As it was shown in Section 1.3 (see also references 3–4), for highly absorbing materials irradiated by a laser of constant intensity, the stationary regime of evaporation is reached within the time on the order of $\tau_0 \approx \chi/V_s^2$. Transforming (5.18) to dimensional variables, we obtain:

$$\tilde{\gamma}_m \approx (V_s^2/\chi)(M/ck_B)\left[L_0/T_s(0)\right]^2 \gg \tau_0^{-1}$$

Thus, when the laser intensity is significantly above the threshold and Equation (5.18) holds, the growing perturbations could destroy the plane phase boundary before the stationary mode of evaporation is established. In this regard, a question arises about how perturbations of a plane vaporization front approaching a stationary mode evolve. To examine this question, we shall now investigate the development of small perturbations of plane nonstationary evaporation front and corresponding temperature field. We shall assume that the laser radiation is switched on at time $t = 0$, and has a constant intensity I, which considerably exceeds the instability threshold of a plane stationary wave. From the previous discussion we may expect that there are two different types of evaporation front motion: "slow" one-dimensional motion, associated with a change of characteristics of the plane vaporization wave, and "fast" motion, associated with the development of non-one-dimensional perturbations. This enables us to investigate the problem in an adiabatic approximation, i.e., to study the stability of one-dimensional motion with respect to perturbations of the form:

$$\delta\Theta = f(\zeta)\exp\left[ik\eta + \int_0^\tau \gamma(\tau')\,d\tau'\right] \qquad \delta Z = \Xi \exp\left[ik\eta + \int_0^\tau \gamma(\tau')\,d\tau'\right]$$

assuming that $\gamma(\tau)$ is a slowly varying function of time:

$$|d\gamma/d\tau| \ll \gamma^2$$

The last condition means that the change in the instability growth rate in the characteristic time of the instability's development should be small.

It is convenient to consider the instability problem in a coordinate frame connected with an unperturbed evaporation front. Since in the present case the front velocity changes with time, we must slightly modify our definition of dimensionless variables. We assume now

$$\zeta = \mu \left[z - \int_0^t V(t')\, dt' \right] \qquad \delta Z = \mu \left[Z(y,t) - \int_0^t V(t')\, dt' \right]$$

The quantity χ/V_s^2 is no longer a characteristic constant, so that the dimensionless time τ is now chosen in the form

$$\tau = t\chi\mu^2$$

The remaining notations agree with those used previously in this chapter.

The analysis, which is similar to that described above, leads to the dispersion equation in the form:

$$\gamma = \beta \frac{\dot{\Theta}_0 + (\alpha - \beta)\left[\lambda\beta - (\alpha - \beta + 1)^{-1} - \alpha\Lambda(\alpha\Theta_0 + \lambda\beta)\right]}{\alpha\Theta_0(\Theta_0 p^{-1} - \Lambda\beta) + \lambda\beta(1 - \Lambda\beta)} \qquad (5.20)$$

$$k^2 = \alpha(\alpha - \beta) - \gamma \qquad \operatorname{Re}\alpha > 0$$

Here $\Theta_0 = \Theta_0(\tau)$ is the temperature on the surface of the condensed phase, the dot indicates a derivative with respect to time, and $\beta = \beta(\tau) = \mu V(t)/\chi$, where $V(t)$ is the instantaneous velocity of the unperturbed plane evaporation front. The asymptotic values of the front temperature and velocity as $\tau \to \infty$ (corresponding to the stationary evaporation wave) will be marked by the additional subscript s, i.e., they will be denoted as Θ_{0s} and β_s, respectively. Note that γ in Equations (5.11) and (5.20) differs by the factor β_s^2 due to the difference in the definition of dimensionless time.

Despite the formal resemblance between the dispersion equations (5.11) and (5.20), the spectrum of perturbations, which develop on the background of a nonstationary wave differs substantially from the corresponding spectrum in the stationary problem. The difference is because $\beta(\tau)$ in (5.20) varies with time: $\beta(0) = 0$ and $\beta(\tau) \to \beta_s$ as $\tau \to \infty$. Since λ is independent of τ, the second term in the numerator of Equation (5.20) is negative at small τ. This means that $\gamma(k)$ decreases monotonically as k increases. Thus, the plane nonstationary evaporation wave should be stable at significantly small τ for any value of laser radiation intensity. Since $\beta(\tau)$ increases with time, the contribution of the term $\lambda\beta$ in the numerator of Equation (5.20) increases and at a certain time $\tau = \tau_0$ the growth rate $\gamma(k)$ becomes positive, first at some point $k = k_0 \neq 0$. Afterward, the instability region expands to a finite interval and the value of γ continues to grow. As an example, a family of dispersion

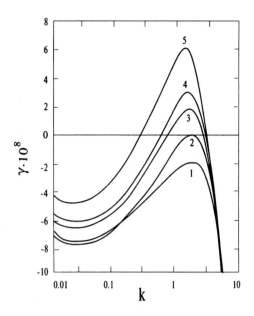

Figure 5.3: Dispersion curves for a nonstationary evaporation wave. Laser intensity $I = 10^{-4}\, \rho V_0 L_v$, $\Lambda = 10^{-5}$.
1— $\tau = 10^{-7}$, $\beta = 3.4 \times 10^{-5}$, $\Gamma = -0.11$,
2— $\tau = \tau_0 = 1.7 \times 10^{-7}$, $\beta = 4.6 \times 10^{-5}$, $\Gamma = -0.15$,
3— $\tau = 2.5\tau_0$, $\beta = 6.2 \times 10^{-5}$, $\Gamma = 0.35$,
4— $\tau = 3.8\tau_0$, $\beta = 6.6 \times 10^{-5}$, $\Gamma = 1.10$,
5—stationary regime, $\beta = \beta_s = 7.5 \times 10^{-5}$.

curves is shown in Figure 5.3 for different times ($\tau_1 < \tau_0$, $\tau_2 = \tau_0$, $\tau_3 > \tau_0$) along with the curve corresponding to the stationary evaporation wave $\tau \to \infty$. Note that the function $\gamma(k)$ reaches its maximum when $\alpha \gg \beta$. Hence, for the position of the maximum k_m, Equation (5.16) holds $k_m \approx (2\Lambda\Theta_0(\tau))^{-1/3}$. This means that the maximum of the growth rate is displaced with time. The growth of perturbations with time may be characterized by the integral

$$\Gamma(\tau) = \max_k \left[\int_0^\tau \gamma(\tau')\, d\tau' \right]$$

For curves 1, 2, and 3 in Figure 5.3, the values of Γ are -0.11, -0.15, and 0.35, respectively. We may estimate the value of τ_0 from the condition:

$$\max_k [\gamma(k, \tau_0)] = 0$$

Using Equation (5.17), we obtain the following equation for τ_0:

$$\lambda \beta(\tau_0) \approx [\Lambda \Theta_0(\tau_0)]^{1/3} \tag{5.21}$$

To determine τ_0 explicitly, we should know the dependence $\beta(\tau)$. This dependence was found using a numerical solution of the heat conduction problem (1.17) and (1.18) (see references 5–6). To obtain an order-of-magnitude estimate of τ_0, we use the simplest analytical approximation for $\beta(\tau)$, which is in reasonable agreement with the numerical calculations

$$\beta(\tau) = \beta_s \exp(-1/\tau\beta_s^2) \tag{5.22}$$

Substituting Equation (5.22) into (5.21) and assuming that $\lambda\beta \approx 1$, we obtain for τ_0:

$$\tau_0 \approx -\frac{3}{\beta_s^2 \log(\Lambda\Theta_{0s})}$$

The characteristic time τ^* of the destruction of the plane nonstationary evaporation wave can be determined from the condition

$$\Gamma(\tau^*) = \max_k \left[\int_0^{\tau^*} \gamma(\tau, k)\, d\tau\right] \approx 1$$

It is easy to see that due to the rapid increase in $\gamma(\tau, k)$ when $\tau > \tau_0$, τ^* must be close to τ_0. Thus, despite the fact that the laser intensity I considerably exceeds the threshold intensity I^*, the time required for destruction of the plane nonstationary wave is of the same order of magnitude as the time to form the stationary evaporation wave.

5.3 Instability of plane stationary evaporation waves in dielectrics

In this section we shall study the instability of laser vaporization of materials whose absorption coefficient is strongly temperature-dependent.[7] Laser breakdown and vaporization of nonlinearly absorbing materials have been considered in Chapter 2. It was established that to maintain the vaporization process, a laser intensity well below the optical breakdown threshold is required. To initiate a vaporization wave at this intensity, an absorbing "seed" is necessary, which ensures a local temperature increase and creation of an absorbing layer with a sufficiently high free electron concentration. A qualitatively similar situation occurs in gases,[8] in which the ionization waves supported by laser radiation may propagate at laser intensities considerably below the breakdown threshold of a cold gas.

Now, we shall recall some of the properties of vaporization waves in transparent dielectrics. First, the calculations[9,10] show that the optical absorption in such waves is small, $A = I_{abs}/I_0 \ll 1$, decreasing with increasing laser intensity. In other words, a considerable fraction of the incident light passes through the wave not being absorbed. Second, the geometric thickness of the absorbing layer in the dielectric is much less than the characteristic scale of the temperature distribution. To simplify the instability analysis, we shall neglect a small change in the light intensity inside the absorbing layer. The

CHAPTER 5. INSTABILITY OF SUBLIMATION WAVES

calculation of the evaporation wave parameters is reduced, in this case, to determining the temperature field inside the dielectric.

As mentioned in Chapter 2, the evaporation waves were observed experimentally on both the incident and exit surfaces of the target. In the limiting case of very small absorption, the both experimental situations are described by the same equations.

Consider a uniform laser beam with constant intensity, I_0, travelling through a dielectric along the z-axis. The temperature of the dielectric satisfies the heat conduction equation:

$$c\rho \partial T/\partial t = \text{div}\,(\kappa \nabla T) + \mu(T) I_0 \qquad (5.23)$$

with the boundary conditions

$$\kappa \nabla T = \rho L_v V_n(T) \quad \text{and} \quad V_n = V_0 \exp(-U/T) \qquad (5.24)$$

on the vaporization front $z = Z(y,t)$, and

$$T \to 0 \quad \text{at } z \to \infty \qquad (5.25)$$

As in the previous section, we assume that the latent heat of vaporization depends on the curvature of the boundary and is given by $L_v = L_0 + \sigma/\rho R$, where R is the local radius of curvature of the boundary. We consider vaporization into a vacuum, and the vapor is assumed to be transparent for the laser light. For the temperature dependence $\mu(T)$ of the absorption coefficient of the dielectric, the same expression is used as in Chapter 2. This choice is justified because the experimental dependence of the stationary vaporization wave parameters on the laser light intensity is in satisfactory agreement with calculations described in Chapter 2 (for a constant thermal conductivity, see references 9 and 10).

The boundary value problem in Equations (5.23) through (5.25) has a particular solution describing a planar stationary evaporation wave: $T(y,z,t) = T_s(\zeta)$, where $\zeta = z - V_s t$. To consider the stability of this solution, let us introduce small perturbations of the phase boundary and temperature field in the form:

$$\begin{aligned} \delta Z(y,t) &= Z(y,t) - V_s t = \Xi \exp(\gamma t + iky) \\ \delta T(\zeta,y,t) &= T(z,y,t) - T_s(\zeta) = f(\zeta) \exp(\gamma t + iky) \end{aligned} \qquad (5.26)$$

Substituting (5.26) into (5.23) and confining ourselves to linear terms in the small quantity $f(\zeta)$, we find that

$$\chi f'' + V_s f' + f[\chi U(\zeta) - \gamma - \chi k^2] = 0 \qquad (5.27)$$

Here

$$U(\zeta) = (I_0/\kappa) d\mu/dT|_{T=T_s}$$

and $\chi = \kappa/c\rho$ is the thermal diffusivity. As in the previous section, we have to project the boundary condition (5.24) related to the vaporization front $z = Z(y,t)$ onto the $\zeta = 0$ plane. Having expanded Equation (5.24) in the power series in terms of the small perturbations δZ and δT, we obtain in the linear approximation:

$$f'(0) + \alpha f(0) = 0 \tag{5.28}$$

where

$$\alpha = \frac{V_s}{\chi} \frac{V_s^2 + \gamma\chi + \Lambda(k\chi)^2}{V_s^2 - m\epsilon[\epsilon\gamma\chi + \Lambda(k\chi)^2]} \tag{5.29}$$

$$\Lambda = \sigma V_s/\rho L_0 \chi \qquad \epsilon = k_B T_s(0)/L_0 \qquad m = cM/k_B$$

k_B and M are the Boltzmann constant and molecular mass, respectively. We note that unlike the case of linear absorption considered in Section 5.1 the perturbation of phase boundary δZ does not appear in the equation for the temperature perturbation, thus the analysis of stability is somewhat simplified.

The instability growth rate $\gamma(k)$ is the eigenvalue of the problem (5.27) and (5.28). It is convenient to transform Equation (5.27) into the Schrödinger equation. Setting $f(\zeta) = \phi(\zeta)\exp(-\zeta V_s/2\chi)$ in (5.27), we find

$$\phi'' + (\epsilon - U)\phi = 0 \tag{5.30}$$

where $\epsilon = k^2 + \gamma/\chi - (V_s/2\chi)^2$. Boundary condition (5.28) then becomes

$$\phi'(0) + \omega\phi(0) = 0 \qquad \omega = \alpha - V_s/2\chi$$

The temperature derivative of the absorption coefficient is analogous to the potential term in the Schrödinger equation (5.30). The basic temperature dependence of $\mu(T)$ is described by the factor $\exp(-E/T)$, where $E \gg T$. It is easily seen that because of the strong temperature dependence of $\mu(T)$, the function $U(\zeta)$ in (5.30) is nonzero only in a narrow interval close to the maximum of the unperturbed temperature $T_s(\zeta)$. The determination of the eigenvalue $\gamma(k)$ is reduced, in this case, to solving a standard problem (well known in quantum mechanics) on the energy level in a short-range potential field (a slight difference is in the different boundary condition). Straightforward calculations lead to the following system of equations for the instability growth rate $\gamma(k)$:

$$2p(p-\omega) = J[p - \omega + (p+\omega)\exp(-2p\zeta_0)]$$
$$p^2 = (V_s/2\chi)^2 + \gamma/\chi + k^2 \quad \omega = \alpha - V_s/2\chi \quad \mathrm{Re}\, p > V_s/2\chi \tag{5.31}$$

where

$$J = (I_0/\chi) \int_0^\infty \mu'(T_s)d\zeta$$

ζ_0 is the coordinate of the maximum of the function $T_s(\zeta)$, and α is given by (5.29). The set of equations (5.31) can be reduced to a single equation for the parameter p,

$$(2p - J)S_- + JS_+ \exp(-2p\zeta_0) = 0 \tag{5.32}$$

where

$$S_\pm = (p \pm V_s/2\chi)^2 - (1-\Lambda)k^2 + m\epsilon^2(1/2 \mp p\chi/V_s)S_0$$
$$S_0 = p^2 - k^2(1 - \Lambda/\epsilon) - (V_s/2\chi)^2$$

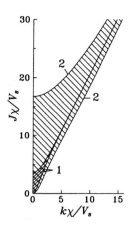

Figure 5.4: The regions of oscillatory (shaded) and aperiodic perturbation growth. 1—$\zeta_0 V_s/\chi = 1$, 2—$\zeta_0 V_s/\chi = 0.1$.

In terms of the solution of Equation (5.32), the instability growth rate is given by
$$\gamma = \chi(p^2 - k^2) - V_s^2/4\chi.$$
The inequalities $\Lambda \ll m\epsilon^2 \ll 1$ and ζ_0 being on the order of, or smaller than, χ/V_s, which follow from an analysis of the unperturbed solution[9,10] are used to solve Equation (5.32).

It is easy to see that the range of parameters, within which the growth rate $\gamma(k)$ has a nonzero imaginary part (oscillating solutions), coincides with the range of parameters, in which Equation (5.32) has complex roots. This region is shown in the (J, k)-plane in Figure 5.4. The location of the boundaries of this region depends on the parameter ζ_0.

In the study of instability, we are interested in modes with $\mathrm{Re}\,\gamma > 0$, which correspond to the solutions of Equation (5.32) with a sufficiently large real part, $\mathrm{Re}\,p > \mathrm{Im}\,p$, k, and V_s/χ. Consider, first, the limiting case $\mathrm{Re}\,(p\zeta_0) \gg 1$, when the second term in (5.32) gives a small correction. Neglecting this term, we obtain two equations:
$$2p - J = 0 \qquad S_- = 0$$

One of the branches of the solution is, thus, given by

$$p_1 = (J/2)(1 + R_1) \qquad (5.33)$$

where
$$R_1 \approx (-1 + m\epsilon^2 \chi J/V_s) \exp(-J\zeta_0)$$

The solution (5.33) exists when $\mathrm{Re}\,(p\zeta_0) \gg 1$ and $R_1 \ll 1$. The last condition is not fulfilled if p_1 is close to the solution of the equation $S_- = 0$. The growth rate at $p = p_1$ is given by

$$\gamma(k) = \chi(J^2/4 - k^2) - V_s^2/4\chi \qquad (5.34)$$

Perturbations with $k = 0$ have the highest growth rate; as k increases, the perturbations become stabilized due to heat conduction. The mode (5.34) is similar to a thermal explosion or optical breakdown initiated by a microinclusion (see Chapter 2).

We note that the wave number k does not appear in the approximate equation (5.33). However, it is easily seen directly from Equation (5.32) that the solution (5.33) is inapplicable in the exponentially narrow region defined by the inequality

$$|(J - V_s/\chi)^2 - 4(1-\Lambda)k^2 + m\epsilon^2(\chi J/V_s + 1)$$
$$\times [J^2 - 4(1 - \Lambda/\epsilon)k^2 - (V_s/\chi)^2]| \leq (4JV_s/\chi)\exp(-J\zeta_0)$$

In this region the growth rate $\gamma(k)$ assumes complex values.

The other branch of the dispersion equation for $J\zeta_0 \gg 1$ describes unstable modes similar to the ones studied in Section 5.1 (corrugation modes). One can find from Equation (5.32), $p_2 = p_{20}(1 + R_2)$, where p_{20} satisfies the equation $S_-(p_{20}) = 0$. For solving this equation, it is convenient to transform S_- into the following form:

$$S_- = (1 + \Delta)$$
$$\times \left\{ \left[p - \frac{V_s}{2\chi(1+\Delta)} \right]^2 - k^2 \left[1 - \frac{\Lambda(1+\Lambda/\epsilon)}{(1+\Delta)} \right] - \left[\frac{V_s \Delta}{2\chi(1+\Delta)} \right]^2 \right\}$$

where $\Delta = m\epsilon^2(1/2 + p\chi/V_s)$. Taking into account the inequality $m\epsilon^2 \ll 1$ the solution of the equation $S_- = 0$ can be written as:

$$p_{20} = V_s/2\chi(1+\Delta) + k[1 - \Lambda(1+\Delta/\epsilon)/(1+\Delta)]^{1/2} + R_{20}, \qquad (5.35)$$

where the correction

$$R_{20} \approx [V_s\Delta/2\chi(1+\Delta)]^2/[p - p_{20} - V_s/2\chi(1+\Delta)]$$

is always small. In limiting cases of small and large wave numbers k one can obtain from (5.35)

$$p_{20} = \begin{cases} V_s/2\chi + k(1-\Lambda)^{1/2} & \text{for } k \ll V_s/m\epsilon^2\chi \\ k(1-\Lambda/\epsilon)^{1/2} & \text{for } k \gg V_s/m\epsilon^2\chi \end{cases}$$

The condition $\operatorname{Re} p\zeta_0 \gg 1$ is fulfilled for this branch if $k\zeta_0 \gg 1$. The correction R_2 is approximately:

$$R_2 \approx \frac{4JV_s[V_s^2 - m\epsilon(1-\epsilon)\Lambda\chi^2 k^2]}{(1-2p_{20})\chi k(V_s + 2m\epsilon^2\chi k)^2} \exp(-2p_{20}\zeta_0)$$

Neglecting the corrections R_{20} and R_2, as well as $\Lambda \ll 1$, we obtain from (5.35) the following expression for $\gamma(k)$

$$\gamma(k) \approx \frac{kV_s}{(1+\Delta)} + \Delta(2+\Delta)\left[\frac{V_s}{2\chi(1+\Delta)}\right]^2 - \Lambda\chi k^2 \frac{(1+\Delta/\epsilon)}{(1+\Delta)} \qquad (5.36)$$

CHAPTER 5. INSTABILITY OF SUBLIMATION WAVES

with $\Delta \approx m\epsilon^2 \chi k/V_s$. At small and large values of k we have

$$\gamma(k) = \begin{cases} kV_s - \Lambda\chi k^2 & \text{for } 1/\zeta_0 \ll k \ll V_0/m\epsilon^2\chi, \\ V_s^2/m\epsilon^2\chi - \Lambda\chi k^2/\epsilon & \text{for } k \gg V_s/m\epsilon^2\chi. \end{cases}$$

The dispersion equation (5.36) is similar to that obtained in the case of strongly absorbing materials (see Section 5.1). For this branch, $\gamma(k)$ has a maximum at finite $k \neq 0$. As in sections 5.1 and 5.2, the dependence of the heat of vaporization on the curvature of the phase boundary results in stabilization of the short-wavelength perturbations.

Approximate formulas (5.33), (5.34), and (5.36) are not valid in the small range of $k \approx J/2$, where the two branches of dispersion equation are close to one another. Consider this region, first, in the case $\operatorname{Re} p\zeta_0 \gg 1$. The correction proportional to $\exp(-2p\zeta_0)$ is important in this case only in the narrow range of wave numbers $|k - k_0| \ll V_s/\chi$, where k_0 is determined from the equation $p_{20}(k_0) = J/2$. In this range, an approximate solution of Equation (5.32) can be obtained in the form:

$$p \simeq (2p_{20} - J/4)$$
$$\pm \sqrt{\left(\frac{2p_{20} - J}{4}\right)^2 - \frac{JV_s[4V_s^2 - m\epsilon(1-\epsilon)\Lambda\chi^2 J^2]}{2\chi[V_s^2 + m\epsilon^2\chi J]^2} \exp(-J\zeta_0)}$$

It is easily seen that p and γ are complex when k is close to k_0. When $J\zeta_0 \gg 1$, we have: $\operatorname{Re}\gamma \gg \operatorname{Im}\gamma$, i.e., the steady-state solution is unstable and the instability is oscillatory. The dispersion dependence $\gamma(k)$ is shown in Figure 5.5. For both branches of the dispersion equation (5.32) considered above, $\operatorname{Re}\gamma$ is positive in certain ranges of k. The maximum growth rate has the thermal explosion mode at $k = 0$.

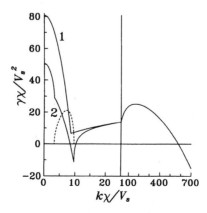

Figure 5.5: The dispersion dependence $\gamma(k)$. $m = 3$, $\epsilon = 0.1$, $\Lambda = 10^{-5}$, $J = 18V_s/\chi$. 1—$\zeta_0 = \chi/V_s$, 2—$\zeta_0 = 0.1\chi/V_s$.

We now consider the case when $J\zeta_0 \ll 1$. Solutions of the thermal-explosion type, described by Equation (5.34), no longer exist. Aperiodic modes of the type (5.36) now are present only for $k\zeta_0 \gg 1$. No solutions with $\text{Re}(p\zeta_0) \gg 1$ exist when $k\zeta_0 \ll 1$.

Consider the solutions satisfying the condition $|p\zeta_0| \ll 1$. Using the series expansion for the exponential function, we find (for $k\zeta_0 \ll 1$, $J\zeta_0 \ll 1$):

$$p = (1 - 2J\zeta_0)V_s/2\chi \pm \left[k^2 - V_s J(1 + J\zeta_0)/\chi\right]^{1/2} \tag{5.37}$$

For small k, Equation (5.37) gives complex p, which correspond to oscillating perturbations with growth rate

$$\gamma(k) = -V_s J \pm V_s(k^2 - V_s J/\chi)^{1/2} \tag{5.38}$$

Solutions described by (5.37) with $|p\zeta_0| \ll 1$ also can be constructed for the values $J\zeta_0 \gg 1$, $k\zeta_0 \gg 1$. Note that perturbations with $k\zeta_0 > 1$ and $|p\zeta_0| < 1$ clearly have $\text{Re}\,\gamma < 0$, i.e., they decay. Perturbations, for which $|\text{Im}\,p| \gg \text{Re}\,p$, also are damped and not considered here.

5.4 Stability analysis of plane stationary ablation waves in polymers

In Section 2.3, we studied the model of UV-laser ablation of organic polymers. The model describes the possibility of "cold" ablation, i.e., it yields realistic ablation rates at moderate surface temperatures, which are in agreement with many experimental observations. It was shown in Section 2.3 that the temperature distribution in the stationary ablation wave has a maximum behind the ablation front. Similar maxima appear also in laser evaporation of metals and transparent dielectrics and can lead to the corrugation instability of vaporization front. In this section the stability of the plane stationary ablation wave will be studied. It will be demonstrated that in polymer ablation, excited species stabilize the ablation front when the thermal relaxation times are of the order of $\tau_T \geq 10^{-10}$ s.

We shall describe the ablation process by the equations for the number density of excited particles N^*, the laser intensity, and the temperature, T. These equations can be written as:

$$\partial N^*/\partial t = V \partial N^*/\partial z + (\sigma I/h\nu)(N_0 - 2N^*) - N^*/\tau_T \tag{5.39}$$
$$\partial I/\partial z = -\sigma I(N_0 - 2N^*) \tag{5.40}$$
$$\partial T/\partial t = V \partial T/\partial z + \chi(\partial^2 T/\partial z^2 + \partial^2 T/\partial y^2) + (\chi/\kappa)Q \tag{5.41}$$

Here we use the same notations and scaling values as in Chapter 2.

We now investigate the stability of the stationary solutions $N_s^*(z)$, $I_s(z)$, $T_s(z)$ of Equations (5.39) through (5.41). We consider the perturbations of the ablation front $\delta Z(y,t)$ shown in Figure 5.1. The boundary conditions on the perturbed ablation front are written in the form:

$$\kappa \nabla_n T|_{z=\delta Z(y,t)} = \rho \left[x \Delta H^* \Pi^* + (1-x) \Delta H \Pi \right]|_{z=\delta Z(y,t)} \quad (5.42)$$

where $\kappa \nabla_n T$ is the normal flux component for the perturbed ablation front, $x(z,y,t) = N^*(z,y,t)/N_0$, $\Pi^* = V_0^* \exp[-E^*/T \delta Z(y,t)]$, $\Pi = V_0 \exp[-E/T \delta Z(y,t)]$, and ΔH^* and ΔH are the enthalpies of vaporization of ground-state and excited-state species, respectively. The velocity of the perturbed ablation front is given by:

$$V = V_s + \partial \delta Z / \partial t = [x \Pi^* + (1-x) \Pi]|_{z=\delta Z(y,t)}$$

Note that the activation energies E and E^* are assumed now to be dependent on the local curvature of ablation front. We shall use the same approximation as in Sections 5.1 through 5.3: $E = E_0 - \sigma M / \rho k_B R$, and similar expression for E^*, where σ is the surface tension (surface energy density), M is the molecular mass, and R is the local radius of curvature of the perturbed ablation front. We assume for simplicity that σ does not depend on the concentration of excited species.

The perturbed quantities can be written as

$$\begin{aligned}
\delta Z(y,t) &= Z_1 \exp(\gamma t + iky) \\
N^*(z,y,t) &= N_s^*(z) + N_1(z) \exp(\gamma t + iky) \\
T(z,y,t) &= T_s(z) + T_1(z) \exp(\gamma t + iky) \\
I(z,y,t) &= I_s(z) + I_1(z) \exp(\gamma t + iky)
\end{aligned} \quad (5.43)$$

Substituting (5.43) into (5.39) through (5.41) and performing linearization with respect to the small perturbations Z_1, N_1, T_1, and I_1, we obtain a set of ordinary differential equations. The boundary conditions for these equations can be obtained from the exact boundary condition on the ablation front $z = \delta Z(y,t)$ employing the standard procedure of expansion of (5.42) in series in the small quantity δZ. The instability growth rate $\gamma(k)$ is an eigenvalue of the resulting boundary value problem. The lengthy but straightforward calculations of the dispersion equation $\gamma(k)$ were performed by Luk'yanchuk et al.[11] We shall now consider the results of these calculations.

The dependence $\gamma(k)$ is shown in Figure 5.6 for $I_0 = I_b = 10^7$ W/cm^2 and for various values of τ_T. The time unit used for the normalization of γ and τ_T is $t_0 = 10^{-8}$ s. The parameters employed in the calculation of $\gamma(k)$ are typical for organic polymers. The figure shows that the dependence $\gamma(k)$ for the polymer ablation is qualitatively similar to that obtained in Section 5.1 for the evaporation of metals. In both cases, a finite region of wave numbers exists where $\gamma(k)$ is positive and the perturbations increase with time. However, the instability appears only when the relaxation process is rather fast. For the parameters employed in Figure 5.6 the critical value of relaxation time is $\tau_T = 20$ ps. When the relaxation time is longer, the ablation front becomes stable. The critical relaxation time depends on the laser intensity. In Figure 5.7 the region of parameters I_0/I_b and τ_T/t_0 is shown where the instability can occur.

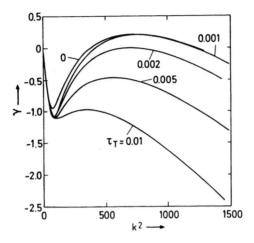

Figure 5.6: The dispersion dependence for different values of relaxation time (in units 10^{-8} s). The laser intensity $I_0 = I_b = 10^7$ W/cm^2.

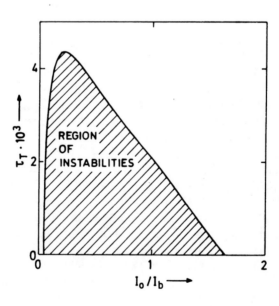

Figure 5.7: The region of instability (shaded) in the (I_0, τ_T)-plane.

It is seen that for relaxation times longer than the critical time $\tau_T^* \simeq 4\times 10^{-11}$ s the ablation front is stable at any laser intensity. For different polymers the critical relaxation time is between 10^{-10} and 10^{-11} s. So we consider $\tau_T > 10^{-10}$ s as a sufficient condition for stability of the ablation front.

Now we shall discuss the physical mechanism of stabilization. Returning to Equation (5.19), we see that maximum instability growth rate is $\gamma_m \simeq dV_s/dz$. It is positive when the temperature grows with z. Considering the ablation model proposed in reference[11], the ablation velocity is a function of both temperature and concentration of the excited species. Thus, the instability criterion is:

$$dV_s/dz|_{z=0} = (\partial V_s/\partial T)(dT_s/dz) + (\partial V_s/\partial N^*)(dN_s^*/dz) > 0 \qquad (5.44)$$

If $E^* \ll E$ the derivative $\partial V_s/\partial N^* > 0$, and the second term in Equation (5.44) becomes negative because $dN_s^*/dz < 0$ (see Chapter 2, Figure 2.10). Thus, this term that originates from the activated desorption of excited species, stabilizes the ablation front.

From a physical point of view, the suppression of the ablation front instabilities can be understood as follows:[11] with increasing relaxation times the contribution of excited species to the total ablation rate increases. Hence, the effective activation energy and, thereby, the value of dV_s/dT decreases. Additionally, as is seen from (5.44), the dependence of the ablation velocity on the concentration of excited species leads to stabilization of the ablation front.

REFERENCES

1. Anisimov, S. I., Tribelskii, M. I. and Epelbaum, Ya. G., Instability of plane evaporation front in laser interaction with a medium, *Sov. Phys.-JETP*, **51**, 802, 1980

2. Gol'berg, S. M. and Tribel'skii, M. I., Development of instability of the transient evaporation of condensed media subjected to radiation, *Sov. Phys.-Solid State*, **24**, 444, 1982

3. Anisimov, S. I., Evaporation of a light-absorbing metal, *High Temperature*, **6**, 110, 1968

4. Anisimov, S. I., Imas, Ya. A., Romanov, G. S., and Khodyko, Yu. V., *Action of High-Power Radiation on Metals*, National Technical Information Service, Springfield, Virginia, 1971

5. Anisimov, S. I., Gol'berg, S. M., Sobol', E. N., and Tribel'skii, M. I., Oscillatory evaporation of condensed media by electromagnetic radiation, *Sov. Tech. Phys. Lett.*, **7**, 379, 1981

6. Gol'berg, S. M., *Instability of Evaporation and Oxidation of Solids Subjected to Laser Radiation*, Candidate of Sci. Dissertation, (Ph. D. Thesis), L. D. Landau Institute for Theoretical Physics, Chernogolovka, 1982

7. Anisimov, S. I. and Khokhlov, V. A., Instability of laser vaporization waves in dielectrics, *Sov. Phys.-Tech. Phys.*, **28**, 765, 1983

8. Raizer, Yu. P., *Laser-Induced Discharge Phenomena*, Consultants Bureau, New York, 1977

9. Aleshin, I. V., Bonch-Bruevich, A. M., Imas, Ya. A., Libenson, M. N., Rubanova, G. M., and Salyadinov, V. S., Laser-induced evaporation of a nonlinearly-absorbing dielectric, *Sov. Phys.-Tech. Phys.*, **22**, 1400, 1977

10. Kovalev, A. A., Laser evaporation of a transparent solid dielectric, *Sov. Phys.-Tech. Phys.*, **24**, 616, 1979

11. Luk'yanchuk, B., Bityurin, N., Anisimov, S., and Bäuerle, D., The role of excited species in UV-laser materials ablation. Part II: The stability of the ablation front, *Appl. Phys.*, **A57**, 449, 1993

Chapter 6

NONLINEAR EVOLUTION OF INSTABILITY IN LASER EVAPORATION

6.1 Slightly supercritical structures in laser sublimation waves

It was shown in Chapter 5 that, under certain conditions, a plane stationary evaporation front can be unstable. The instability occurs when the laser radiation intensity exceeds a threshold which depends on thermophysical properties of the material evaporated. At supercritical laser intensities, a finite interval of wavenumbers $k_{1\gamma} < k < k_{2\gamma}$ exists in which the perturbations of plane front increase exponentially with time. In the weak supercritical regime, i.e., when the laser intensity is above, but close to the threshold, the band of unstable modes is narrow and located near the critical wavenumber k_m, as it is shown in Figure 6.1. Note that this type of instability is observed in crystal growth[1,2] and in laminar flame propagation.[3,4]

We now shall study the nonlinear evolution of unstable modes. A general approach to this problem is based on the numerical solution of heat conduction equation for the condensed phase with appropriate initial and boundary conditions.[5,6] A numerical solution of the three-dimensional heat conduction problem with nonlinear boundary conditions at a moving boundary, whose shape and position were not known beforehand, requires a considerable amount of computer time. The numerical calculations[5,6] are limited to studying the two-dimensional problem in which the temperature field is independent of the x-coordinate. This approach reduces considerably the volume of numerical calculations while preserving some important features of the nonlinear development of the corrugation instability. The equations to be solved

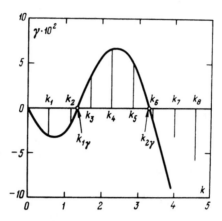

Figure 6.1: Dispersion curve and wavenumbers of the initial perturbations.

are those considered in Chapter 5 (see Equations (5.1) through (5.3)). For convenience, we write them in the dimensionless form as:

$$\partial\theta/\partial\tau = \partial^2\theta/\partial\zeta^2 + \partial^2\theta/\partial\eta^2 + \exp(Z - \zeta) \tag{6.1}$$

$$\partial\theta/\partial\zeta - \partial\theta/\partial\eta \cdot \partial Z/\partial\eta = \lambda \dot{Z}(1 - \Lambda K) \tag{6.2}$$

$$\dot{Z} = \nu \left[1 - (\partial Z/\partial\eta)^2\right]^{1/2} \exp\left[-(1 - \Lambda K)p/\theta\right]$$

where

$$\theta = \mu\kappa T/AI \quad \zeta = \mu z \quad \eta = \mu y \quad \lambda = \mu\kappa L_v/cAI$$
$$\tau = t/\chi\mu^2 \quad \nu = V_0/\mu\chi \quad \Lambda = \sigma\mu/\rho L_v \quad K = 1/\mu R$$

R is the radius of curvature of the perturbed evaporation front and the function $\zeta = Z(\eta, \tau)$ defines the position of the phase boundary. The dimensionless curvature of the phase boundary, K is given by:

$$K = (\partial^2 Z/\partial\eta^2)\left[1 + (\partial Z/\partial\eta)^2\right]^{-1/2} \tag{6.3}$$

The problem (6.1) through (6.3) is best investigated numerically in a laboratory framework, rather than in the framework associated with the vaporization wave front. We consider periodic perturbations of the temperature and phase boundary shape:

$$\theta(\eta, \zeta, \tau) = \theta(\eta + 2\pi/k, \zeta, \tau) \quad Z(\eta, \tau) = Z(\eta + 2\pi/k, \tau)$$

where k is the perturbation wavenumber. The initial conditions to Equations (6.1) through (6.3) are chosen as:

$$\theta(\eta, \zeta, 0) = 0 \quad \text{and} \quad Z(\eta, 0) = Z_0(\eta) \tag{6.4}$$

The problem formulated above was solved by numerical methods.[5-7] The details of the numerical procedure can be found in Gol'berg.[7] Here we will consider the results. First, we examine the situation corresponding to a slightly supercritical laser intensity: $I > I^*$ and $I - I^* \ll I^*$. For this case, a small interval of wavenumbers, $k_{1\gamma} < k < k_{2\gamma}$, $k_{2\gamma} - k_{1\gamma} \ll k_{1\gamma}$, exists in which the linear growth rate $\gamma(k)$ is positive. The computations[5] show that when the initial conditions have the form of "white noise" i.e., the function $Z_0(\eta)$ contains a large number of Fourier harmonics of approximately equal amplitudes, then the nonlinear evolution of instability is universal, regardless of the specific initial shape of the phase boundary. At the initial stage, when the mode amplitudes are small, the growth of each mode is determined by the linear dispersion equation (5.11); with increasing time, the amplitudes of the unstable modes (with $k_{1\gamma} < k < k_{2\gamma}$) increase exponentially, and the amplitudes of the modes lying outside the instability interval $(k_{1\gamma}, k_{2\gamma})$ decrease exponentially. As the amplitudes of the unstable modes increase, the resonant excitation of higher harmonics takes place. The amplitudes of these harmonics also increase with time, despite the fact that the wavenumbers of the harmonics lie in the region of linear stability. Since every mode modulates the temperature field with its own spatial period, the interaction between different modes (not associated with resonant conditions $k_i = nk_j$ where n is an integer) leads to the suppression of all modes, except the mode which has the greatest amplitude when nonlinear interaction of modes begins and its resonant satellites.

For the initial perturbations of the type of "white noise" this is a mode with maximum linear growth rate (see Figure 6.1), since it grows faster than any other during the initial linear stage. Note that for low supercriticality, all the unstable modes are concentrated in a small vicinity of k_m. For this reason, two unstable modes cannot satisfy the resonant condition $k_i = nk_j$, and one can expect that only two modes with $k \approx k_m$ and $k \approx 2k_m$ will exist at $\tau \gg 1$. An example of computations illustrating the evolution of perturbations in a slightly supercritical case is presented in Figures 6.1 through 6.3. Figure 6.1 shows the linear growth rate $\gamma(k)$ with the wavenumbers k_n, which satisfy the condition of periodicity. The initial condition for numerical computations was chosen as a superposition of modes with $k = k_n$:

$$Z_0(\eta) = \sum_{n=1}^{N} A_n(0) \sin k_n \eta \qquad (6.5)$$

where $k_n = nk_1$, $N = 10$, and $A_n(0) = 0.0015 = $ constant. In Figure 6.2 the nonlinear evolution of a vaporization front is shown. The front shape may be represented in the form of the Fourier series:

$$Z(\eta, \tau) = A_0(\tau) + \sum_n A_n(\tau) \sin [k_n + \psi_n(\tau)]$$

where the amplitudes and phases of the Fourier components vary with time. The amplitudes A_n are shown in Figure 6.3 as functions of time. We see from the computations that after some transient process a steady state nonplane vaporization wave moving at constant velocity is formed. The fundamental spatial period of the wave is equal to $2\pi/k_4$, where k_4 is the wavenumber of the mode having the largest value of the linear growth rate (see Figure 6.1). From Figures 6.2 and 6.3, one can see that the shape of the front is not

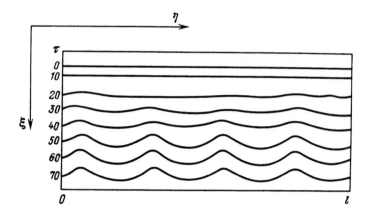

Figure 6.2: The nonlinear evolution of a vaporization front. Initial perturbations of the type of "white noise". Perturbation wavenumbers $k_n = nk_i$. The values of τ are indicated on the left. The distances between successive front positions are reduced by a factor of 10. The scale of the front structure remains unchanged.

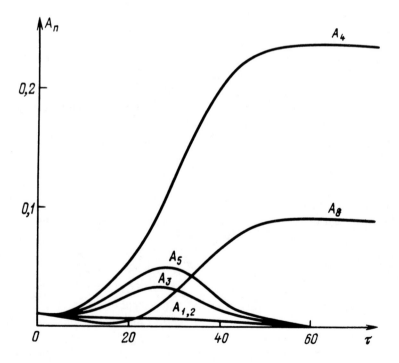

Figure 6.3: Fourier components of the structure shown in Figure 6.2 as functions of time.

sinusoidal. The resonant harmonic having the wavenumber $k_8 = 2k_4$ has a large amplitude—despite the fact that this mode is located inside the linear stability region.

Consider now the case of an almost monochromatic initial condition. Assume that the initial condition (6.5) besides the "white noise" modes contains a mode with $k = k_j$, where k_j is notably different from k_m. The amplitude A_j of this mode is supposed to be much larger than the amplitudes of the "white noise" components A_n. The computations show that, if at the onset of nonlinear interaction between the modes the amplitude A_j is greater than the amplitude A_r of the harmonic whose wavenumber k_r is closest to k_m, which corresponds to the maximum linear growth rate, the interaction of these two modes leads to the formation of a stationary structure with fundamental period $2\pi/k_j$. If, at the nonlinear interaction stage, the amplitude A_j is less than the amplitude A_r of the mode whose linear growth rate is higher, the initial condition is "forgotten" and a structure is formed with a spatial period $2\pi/k_r$. This is illustrated in Figures 6.4 and 6.5, which show the evolution of a vaporization front and the time dependence of the Fourier amplitudes A_n for the initial conditions (6.5) with $A_n = 0.0015$ $(n \neq 3)$, $A_3 = 0.045$.

We see that the structure with the wavenumber k_3 is suppressed at a moment $\tau \approx 50$, despite the fact that its initial amplitude was fairly large and that its growth rate $\gamma(k_3)$ is positive. After a certain transient $(50 < \tau < 70)$, a stationary wave having a period $2\pi/k_4$ is formed. We see, thus, that the structures having a fundamental period $2\pi/k$, whose wavenumbers k lie within the instability interval $(k_{1\gamma}, k_{2\gamma})$ but close to its boundaries, are unstable and disintegrate. To understand this fact, a closed phenomenological equation was proposed[5,8] that describes the shape of the phase boundary in slightly supercritical regime. It was shown that the equation

$$\partial Z/\partial t + \hat{L}Z + (\partial Z/\partial \eta)^2 = 0, \tag{6.6}$$

where $\hat{L}Z = \partial^4 Z/\partial \eta^4 + 2\alpha \partial^2 Z/\partial \eta^2 + Z$, $\alpha = k_m^2 = (\gamma_m + 1)^{1/2} \approx 1 + \gamma_m/2$ and $\gamma_m = \gamma(k_m)$ is a maximum value of $\gamma(k)$ in the instability interval $(k_{1\gamma}, k_{2\gamma})$ (see Figure 6.1), accurately depicts (in two-dimensional case) the results of numerical calculations, allowing a simple investigation of the general properties of unstable evaporation.

Equation (6.6) has stationary solutions in the form:

$$Z(\eta) = A \sin k\eta \qquad A^2 = \alpha^2 - 1 - (k^2 - \alpha)^2 \tag{6.7}$$

where k lies in the interval of instability of the trivial solution $Z(\eta, \tau) = 0$: $[\alpha - (\alpha^2 - 1)^{1/2}]^{1/2} < k < [\alpha + (\alpha^2 - 1)^{1/2}]^{1/2}$. Equation (6.6) was employed for the analysis of stability of steady-state solutions (6.7). The stability criterion can be written in the form:

$$A^2(k) > 2A_m^2/3 \tag{6.8}$$

where $A_m = \gamma_m = (\alpha^2 - 1)^{1/2}$ is the maximum value of the amplitude of stationary solution (6.7). For a given supercriticality, from Equation (6.8), we see that only those stationary structures whose amplitudes are not too small are stable. Those structures whose k lie near the boundaries of the plane front

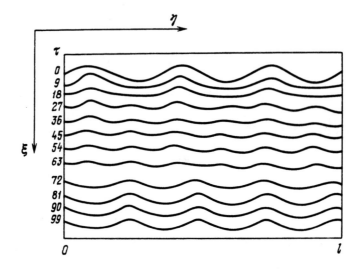

Figure 6.4: The nonlinear evolution of a vaporization front. The initial amplitude of perturbation with $k = k_3$ exceeds the level of "white noise" by a factor of 30.

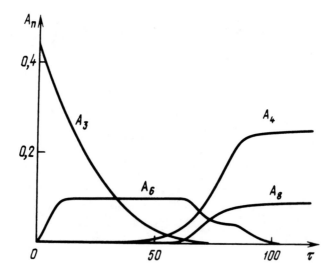

Figure 6.5: Fourier components of the structure shown in Figure 6.4.

CHAPTER 6. NONLINEAR EVOLUTION OF INSTABILITY

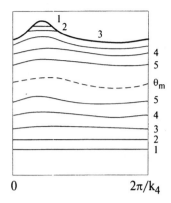

Figure 6.6: Isotherms of the temperature field in the stationary evaporation wave shown in Figure 6.2. Curve 1—$\theta = 0.71\theta_m$, 2—$\theta = 0.78\theta_m$, 3—$\theta = 0.84\theta_m$, 4—$\theta = 0.91\theta_m$, 5—$\theta = 0.97\theta_m$.

instability range are absolutely unstable. Note that an analysis of stability of plane Couette flow leads to a similar criterion.[9] Using condition (6.8), we can conclude that the stationary solution, with fundamental spatial period $2\pi/k_3$, is unstable and should be destroyed at arbitrary initial amplitude A_3.

We now shall discuss the characteristic properties of the stationary structure. The temperature distribution in a nonplane steady-state evaporation wave with fundamental period $2\pi/k_4$ (see Figure 6.1) is shown in Figure 6.6. The phase boundary is not isothermal, and those portions of the boundary displaced to the condensed phase have higher temperatures. Because the evaporation wave is stationary and its propagation velocity is constant, the effective vaporization energy L_{eff} in the "hot" portions of the front must be higher than in the "cold". The curvature of the isotherms decreases quickly with distance from the phase boundary. The isotherm with temperature $\theta = 0.7\theta_m$, where θ_m is the maximum temperature, is almost planar and its position is the same as in the case of the plane vaporization front corresponding to the same laser intensity.

6.2 Strongly supercritical regimes of laser evaporation. "Spin" sublimation of solids.

Previously we have discussed the case of low supercriticality when the interval of unstable modes is narrow and the second harmonic of the perturbed fundamental mode lies in the region of linear stability. In this case, the evolution of the perturbations results in the formation of stationary structures with nonplane, nonisothermal phase boundary. As the supercriticality increases, the wavenumber of the second harmonic approaches the upper boundary $k_{2\gamma}$ of the instability range and the situation may change qualitatively. If the intensity I of the laser radiation greatly exceeds the instability threshold, the range where the linear growth rate is positive can become so wide that multiple harmonics of the monochromatic initial perturbation will lie in it.

This situation occurs when the laser intensity I is greater than I^* determined from the equation $2k_{1\gamma}(I^*) = k_{2\gamma}(I^*)$. Calculations show that in this case an important role is played by the generation of harmonics, which leads to nonstationary regimes of evaporation front propagation. Actually, the steady-state propagation can be violated even earlier, at I somewhat smaller than I^*, since the coupling between the modes can compensate for the small damping of the second harmonic. We refer to Figure 6.1. Suppose that the initial perturbation has the wavenumber k_3, lying in the instability range $(k_{1\gamma}, k_{2\gamma})$. The second harmonics of this perturbation has wavenumber $k_6 = 2k_3$ located in the region of linear stability near the upper boundary $k_{2\gamma}$ of the range of unstable modes. The damping of the second harmonic in this case is small and may be less than the amplification resulting from mode interaction if the amplitude of the fundamental mode $k = k_3$ is fairly large. One might expect a stationary regime of evaporation not to be reached at this level of supercriticality. A numerical computation[6] shows that, after a certain transient, an asymptotic traveling-wave mode is established. We recall that in Anisimov et al.[6] a two-dimensional temperature field was considered. A physical example related to this calculation is the vaporization of a thin-walled cylinder irradiated along its axis. Here, the radial component of the temperature gradient may be neglected, so that the temperature $\theta = \theta(\eta, \zeta, \tau)$ where the η coordinate is measured along the circumference of the base of the cylinder and the ζ axis is parallel to the cylinder axis. The (η, ζ) plane may be considered as the convolution of the side of cylinder. The radius, R, of the cylinder is determined by the condition of periodicity:

$$\theta(\eta + 2\pi R, \zeta, \tau) = \theta(\eta, \zeta, \tau) \qquad \text{or} \quad R = 1/k_1.$$

With this constraint in mind, we now can interpret the results of computations[6] as the formation of the structure which rotates around the cylinder axis and propagates with constant velocity along this axis. The points on the front having the same temperature form helical lines winding along the side of the cylinder. Similar spinning structures were observed in combustion waves.[10] In Figure 6.7 steady-state propagation of the above described "spin" evaporation wave is shown. The steady-state regime is reached after a certain transient. The duration of the transient and the direction of wave motion along the η-axis are determined by the specific initial conditions.

In the Fourier expansion of the phase boundary shape, the modes with wavenumbers k_3 and $k_6 = 2k_3$ are present. The interaction between these modes leads to a periodic variation of their amplitudes. The law of motion of the vaporization front is described by the following equation:

$$Z(\eta, \tau) = \beta\tau + A_3 \sin(k_3\eta - \beta\tau) + A_6 \sin(k_6\eta - \beta\tau + \psi_0)$$

where β, A_3, A_6, and ψ_0 are constants.

CHAPTER 6. NONLINEAR EVOLUTION OF INSTABILITY

Figure 6.7: Steady-state propagation of the "spin" vaporization wave. The displacement of the front is reduced by a factor of 6. The time intervals between subsequent locations of the front $\Delta \tau = 11.4$, $l = 2\pi R = 2\pi/k_3$.

When the laser intensity I reaches I^*, the second harmonics wavenumber falls inside the range of linear instability. Numerical computations show that when this happens, a complex, nonstationary propagation mode arises. An example of this mode is presented in Figure 6.8. Unlike the case of "spin" evaporation, a Fourier series expansion of the function $Z(\eta, \tau)$ in this case shows a nonperiodic time behavior of the front shape. However, to answer the question of whether this mode is complex periodic, quasi-periodic, or true stochastic, however, one needs more information concerning the lasting behavior of Fourier amplitudes. We note that this question seems to be academic, because in experiments it is practically impossible to establish decisively which of the complex, nonstationary vaporization modes occurs.

6.3 Nonlinear stages of instability of evaporation waves in dielectrics

As was shown in Chapter 5, there are two different modes of instability of plane evaporation waves in dielectrics. The first is similar to the corrugation instability of evaporation waves in highly absorbing solids. For this mode the growth of perturbations is aperiodic and the growth rate reaches its maximum at finite k. The second one is the instability mode similar to thermal explosion. Maximum growth rate is reached for this mode at $k = 0$. When $J\xi_0 \gg 1$, there is no solution of the thermal-explosion type. If $k\xi_0 \ll 1$, the growth rate is given by Equation (5.38):

$$\gamma(k) = -JV_s + V_s(k^2 - JV_s/\chi)^{1/2} \qquad (6.9)$$

from which we can see that the perturbations may grow in an oscillatory manner if k is small. A more detailed analysis shows that oscillatory solutions

102 INSTABILITIES IN LASER-MATTER INTERACTION

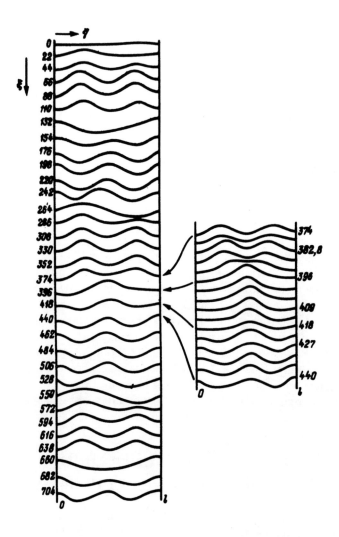

Figure 6.8: The shape of the phase boundary for the "stochastic" vaporization mode. The distances between subsequent locations of the front are reduced by a factor of 20. The values of τ are indicated on the left.

exist for arbitrary J. However, the range of wavenumbers corresponding to these solutions decreases rapidly as J increases.

The region where oscillatory solutions exist is shown cross-hatched in Figure 5.4. It is seen that in a wide range of laser intensities, the oscillatory solutions are possible if ζ_0 is sufficiently small. One may expect that the nonlinear evolution of some modes from this range will lead to the formation of plane oscillatory evaporation waves. This assumption is confirmed by numerical calculations.[11,12] The calculations[11] were performed for a simplified model in which the temperature dependence of thermal conductivity was not considered, and the light absorption was described by the simplest Bouguer law. The results of calculations[11] show that three types of plane vaporization waves exist in nonlinearly absorbing media: (1) stationary waves stable to perturbations with $k = 0$; (2) oscillatory waves; and (3) waves in which vaporization ceases because of bleaching of material. In reference[12], the electronic heat conduction was taken into account and the light absorption was calculated by solving Maxwell's equations for the electromagnetic field of a light wave. We now discuss these calculations more in detail. In this work, as in Chapters 1 and 2, the energy transfer in condensed phase is described by the heat conduction equation

$$c\rho \partial T/\partial t = \partial/\partial z[(\kappa_0 + \kappa_e)\partial T/\partial z] + c\rho V \partial T/\partial z + Q \qquad (6.10)$$

where κ_0 and κ_e are the phonon and electron components of thermal conductivity. The heat source, Q, in Equation (6.10) is given by formula (1.24). To determine the electric field the wave equation (1.25) is solved numerically. The complex dielectric constant of ionized dielectric is written in the form (see, Poyurovskaya et al.[13]):

$$\epsilon = \epsilon_0 - 4\pi i \sigma/\omega \qquad \epsilon_0 = n_0^2 - 4\pi \sigma/\nu \quad \text{and} \quad \sigma = \sigma_0 \nu^2/(\nu^2 + \omega_0^2)$$

where σ is the electrical conductivity at laser frequency ω_0; σ_0 is the static conductivity; ν is the effective collision frequency; and n_0 is the refraction index of the cold dielectric. The thermal and static electrical conductivities are assumed to be proportional to the concentration of free electrons, that is strongly temperature dependent:

$$\kappa_e(T) = \kappa_{e0} \exp(-E_g/T) \quad \text{and} \quad \sigma_0(T) = \sigma_{00} \exp(-E_g/T) \qquad (6.11)$$

The usual boundary conditions (1.18) are imposed at the vaporization front $z = Z(t)$.

Equation (6.10) has a steady-state solution $T_s(z)$. This solution was found[12] by numerically solving Equation (6.10) (with $\partial T/\partial t = 0$) together with the wave equation (1.25). Then, a small perturbation was introduced and its evolution was studied by solving numerically the complete time-dependent equation (6.10). The computations show that in a certain range of laser intensities, the stationary solution $T_s(z)$ is stable. Initial perturbations of the temperature field decrease with time, so that the temperature and the front velocity tend to their stationary values: $T(z,t) \to T_s(z)$, and $V(t) \to V_s$. Outside the stability region small one-dimensional perturbations grow with

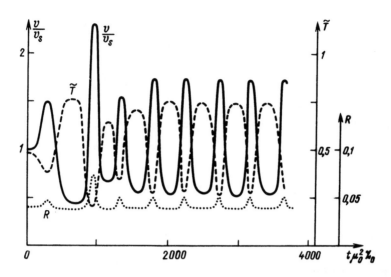

Figure 6.9: The self-excited oscillation mode of dielectric vaporization. Dependence of the front velocity v and the reflection and transmission coefficients, R and $\tilde{T}$, on time.

time. The evolution of the perturbations and the character of the asymptotic propagation mode depend on the laser intensity and parameters of the material irradiated. Three asymptotic regimes were observed in computations:[12] (1) stable nonlinear oscillations with periodic variation of the front velocity (Figure 6.9); (2) low-temperature regime with complete bleaching of a specimen (Figure 6.10); and (3) separation of the absorbing layer from the phase boundary and formation of ionization wave not connected with phase transition (Figure 6.11) Note, that the third regime is possible only when light is incident from the side of cold dielectric.

As is seen in Figure 6.9, the oscillations of vaporization front velocity are accompanied by the oscillations of reflection, R, and transmission, $\tilde{T}$, of laser radiation. The period of oscillations is, in order of magnitude, equal to $\chi(T_{sm})/V_s^2$, where $\chi(T_{sm})$ is the thermal diffusivity at the maximum steady-state temperature, T_{sm}. The radiation is incident on the specimen from the side of vaporization front. A similar regime also exists when the radiation is incident from the side of the cold dielectric. The parameters of the target material in the calculation presented in Figure 6.9 are close to those of fused quartz. For these parameters of the target and the same laser intensity there exists, formally, a stationary solution of Equation (6.10) with temperature profile $T_s(z)$ and front velocity V_s, which is unstable. In the stationary evaporation mode, the light-absorbing layer (i.e., the neighborhood of the point where the temperature reaches its maximum) and the phase boundary move at the same velocity, V_s. In the oscillatory mode, the distance between these regions changes periodically with time. Over some regions of parame-

CHAPTER 6. NONLINEAR EVOLUTION OF INSTABILITY

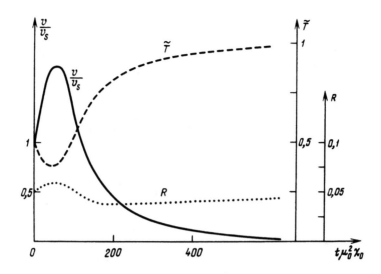

Figure 6.10: Unstable modes of dielectric vaporization. Irradiation from the side of the vaporization front; the front moves more rapidly than the absorbing layer.

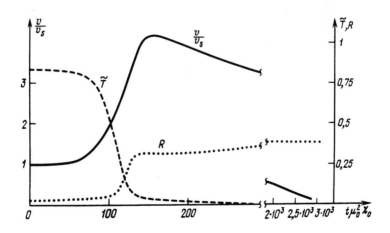

Figure 6.11: Unstable modes of dielectric vaporization. Irradiation from the side of cold dielectric; the absorbing layer moves faster than the vaporization front.

ters, the synchronism in propagation of the absorbing layer and vaporization front may break down. If the vaporization front moves faster, it overtakes and "swallows" the absorbing layer. The optical thickness of the absorbing layer rapidly decreases and the specimen becomes transparent. An example of one-dimensional instability leading to the bleaching of the dielectric target is shown in Figure 6.10 (the laser beam comes in from the side of the vaporization front).

A different mode of instability may arise if the absorbing layer moves faster than the vaporization front. In this case, the absorbing layer breaks away from the phase boundary, and the heat flux to the vaporization front decreases. This leads to the deceleration of the phase boundary and formation of an absorption wave similar to the optical breakdown waves in gases.[13,14] According Anisimov et al.,[12] this instability occurs when the laser radiation is focused on the exit surface of dielectric sample (irradiation from the side of the cold dielectric). An analysis of the numerical results shows that the evolution of the perturbations of a plane stationary evaporation wave is determined by two parameters: (1) the ratio $P = U/E_g$ of characteristic activation energies for evaporation (U) and for high-frequency conductivity (E_g), and (2) the dimensionless laser intensity $Q = I_0/\rho L_0 V_0$. If radiation is incident from the vaporization front side, and the parameter P does not exceed some critical value $P_1 \approx 1$, then as Q increases at constant P, the following changes occur: the stationary evaporation changes into a self-sustained oscillation mode and then to specimen bleaching due to the absorption layer being vaporized. A further increase in the laser intensity leads to a change in propagation modes in the reverse order. If $P > P_1$, a stationary evaporation is stable with respect to perturbations with $k = 0$. When light is incident from the side of the cold dielectric, and $P < P_2 \approx 1$, the sequence of modes as Q increases is: separation of the absorbing layer from the phase boundary, bleaching of the target due to vaporization of the absorbing layer, oscillatory mode, and a stationary evaporation wave.

REFERENCES

1. Langer, J. S., Instabilities and pattern formation in crystal growth, *Rev. Mod. Phys.*, **52**, 1, 1980

2. Mullins, W. W. and Sekerka, R. F., Morphological stability of a particle growing by diffusion of heat flow, *Journ. Appl. Phys.*, **34**, 323, 1964

3. Zel'dovich, Ya. B., Barenblatt, G. I., Librovich, V.B., and Makhviladze, G. M., *Mathematical Theory of Combustion and Explosion*, Nauka, Moscow, 1980

4. Aldushin, A. P., Zel'dovich, Ya. B., and Malomed, B. A., *Phenomenology of Unstable Combustion for Small and Large Lewis Numbers*, Preprint N 596, Space Research Institute, USSR Acad.Sci., Moscow, 1980

5. Anisimov, S. I., Gol'berg, S. M., Malomed, B. A., and Tribel'skii, M. I., Two-dimensional slightly supercritical structures in laser sublimation waves, *Sov. Phys.-Doklady*, **27**, 130, 1982

6. Anisimov, S. I., Gol'berg, S. M., and Tribel'skii, M. I., Spin regimes of sublimation of solids under the action of laser radiation, *Sov. Phys.-JETP*, **55**, 929, 1982

7. Gol'berg, S. M., *Instability of Evaporation and Oxidation of Solids Subjected to Laser Radiation*, Candidate of Sci. Dissertation (Ph.D.Thesis), L.D. Landau Institute for Theoretical Physics, Chernogolovka, 1982

8. Anisimov, S. I. and Tribel'skii, M. I., Instability and spontaneous symmetry breaking in macroscopic laser-matter interaction, *Sov. Sci. A. Phys.*, Vol. **8**, 259, 1987

9. Kogelman, S. and DiPrima, R. C., Stability of spatially periodic supercritical flows in hydrodynamics, *Phys. Fluids*, **13**, 1, 1970

10. Ivleva, T. P., Merzhanov, A. G., and Shkadinskii, K. G., Mathematical model of spin combustion, *Sov. Phys.-Doklady*, **23**, 255, 1978

11. Gol'berg, S. M., Tribel'skii, M. I., and Khokhlov, V. A., *Nonlinear Oscillations in Laser-Induced Evaporation of Dielectrics*, Preprint N 20, Landau Institute for Theoretical Physics, USSR Acad. Sci., Chernogolovka, 1982

12. Anisimov, S. I., Gol'berg, S. M., Tribel'skii, M. I., and Khokhlov, V. A., Laser-induced evaporation of nonlinearly absorbing media. In: *Effect of High-Intensity Energy Fluxes on Materials*, Ed. by Rykalin, N. N., Nauka, Moscow, 1985

13. Poyurovskaya, I. E., Tribel'skii, M. I., and Fisher, V. I., Ionization wave sustained by intense monochromatic radiation, *Sov. Phys.-JETP*, **55**, 1060, 1982

14. Raizer, Yu. P., Breakdown and heating of gases under the influence of a laser beam, *Sov. Phys.-Uspekhi*, **8**, 650, 1966

Chapter 7

INSTABILITIES IN THE LASER INTERACTION WITH LIQUIDS

7.1 Thermocapillary instability of a liquid heated by laser radiation

In this section we will examine the instability of a plane homogeneous liquid film which absorbs laser radiation. The instability arises from the temperature dependence of surface tension $\sigma(T)$. The capillary forces are small in magnitude; therefore, to observe the liquid motion induced by them it is necessary that other forces—in particular, reactive forces which arise during the evaporation of liquid—do not act on the layer. Therefore, we would expect an instability of this type to be observed at relatively low laser radiation intensities in materials whose vapor pressure at melting point is low.

To understand qualitatively how instability arises, we shall examine a liquid film of thickness, H, situated horizontally in a gravitational field and heated from above by laser radiation. We shall consider the radiation to be absorbed in infinitesimally thin surface layer because, as will be seen, long-wave perturbations have the greatest growth rate. For simplicity, we shall assume that the temperature gradient close to the surface is constant in the steady state:

$$dT_0/dz = AI/\kappa = \text{constant} \tag{7.1}$$

Here, I is the laser radiation intensity, and A is the fraction of laser intensity absorbed by the liquid.

It is well known[1] that there are two types of oscillations which can be excited in liquid film under examination: gravitational-capillary and thermocapillary. Excitation of thermocapillary oscillations is associated with the temperature dependence of surface tension. Because the surface tension decreases with temperature,[2] forces will occur on a nonuniformly heated liquid surface that are directed away from the heated portion toward the cold portions and will cause motion in the layer of liquid adjacent to surface. If the

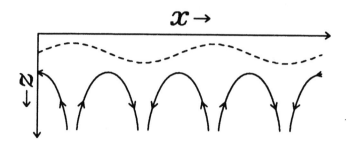

Figure 7.1: Flow structure caused by thermocapillary force. The dashed line represents the temperature distribution; the solid lines represent the flow.

surface temperature profile is periodic, a liquid flow will be set up as shown in Figure 7.1. It can be seen from the flow pattern that liquid in the surface layer will flow from the heated portions to the cold portions and that liquid at a lower temperature will rise from below to replace it. Convective transport and thermal conduction will equalize the temperature and, due to the inertia of the liquid, the equalization process may have an oscillatory nature.

It follows from the arguments presented here that, should the liquid temperature not fall, but increase from below, the thermocapillary effect may lead to an aperiodic instability. Such an instability was studied by Pearson,[3] and will not be considered here.

If the gradient of temperature of the liquid has the sign defined by Equation (7.1), the situation becomes more complex. The interaction of two branches of spectrum of the natural modes—gravitational-capillary and thermocapillary oscillations—may lead to a parametric amplification of the fluctuations, i.e., to the occurrence of instability. For a description of this phenomenon, we shall introduce a scalar potential φ and a vector potential $\mathbf{A}$ of the velocity, satisfying the equations:

$$\mathbf{v} = \nabla\varphi + \operatorname{rot} \mathbf{A} \qquad \operatorname{div} \mathbf{A} = 0$$

where $\mathbf{v}(\mathbf{r},t)$ is the velocity of liquid. Introducing a temperature perturbation $T_1(\mathbf{r},t) = T(\mathbf{r},t) - T_0(z)$, $|T_1| \ll T_0$, and linearizing the Navier-Stokes equations and the heat conduction equation for small values of $\mathbf{v}$ and T_1, we obtain the set of equations:

$$\begin{aligned} \Delta\varphi &= 0 \\ \frac{\partial \mathbf{A}}{\partial t} - \nu\Delta\mathbf{A} &= 0 \\ \frac{\partial T_1}{\partial t} + v_z \frac{dT_0}{dz} &= \chi\Delta T_1 \end{aligned} \qquad (7.2)$$

where $\nu = \eta/\rho$ is the kinematic viscosity and $\chi = \kappa/c\rho$ is the thermal diffusivity of the liquid. The boundary conditions at the lower boundary of the

CHAPTER 7. INSTABILITIES IN LIQUIDS

film, $x = -H$ have the form:

$$T_1 = 0 \quad \text{and} \quad \mathbf{v} = 0 \qquad (7.3)$$

On the upper boundary, the boundary conditions have to be imposed on the perturbed liquid surface $z = Z(x, y, t)$. In such cases, it is convenient to project the boundary conditions onto the plane $z = 0$ by expanding all the functions occurring in the power series in Z. Then, in the linear approximation with respect to T_1, $\mathbf{v}$ and Z, the boundary conditions to the Navier-Stokes equations (with the conditions that all three force components acting on unit area of the surface of a liquid must be equal to zero) have the form:[1]

$$\rho \frac{\partial^2 \varphi}{\partial t^2} - \sigma \left(\frac{\partial^2 v_x}{\partial x^2} + \frac{\partial^2 v_z}{\partial y^2} \right) + \rho g v_z + 2\eta \frac{\partial^2 v_z}{\partial z \partial t} = 0$$

$$\frac{d\sigma}{dT} \frac{\partial}{\partial x} \left(\frac{\partial T_1}{\partial t} + v_z \frac{dT_0}{dz} \right) = \eta \frac{\partial}{\partial t} \left(\frac{\partial v_x}{\partial z} + \frac{\partial v_z}{\partial x} \right) \qquad (7.4)$$

$$\frac{d\sigma}{dT} \frac{\partial}{\partial y} \left(\frac{\partial T_1}{\partial t} + v_z \frac{dT_0}{dz} \right) = \eta \frac{\partial}{\partial t} \left(\frac{\partial v_y}{\partial z} + \frac{\partial v_z}{\partial y} \right)$$

In general, the boundary conditions for the heat conduction equation, that ensures the continuity of normal component of energy flux through the phase boundary, for this problem may be written in form:

$$-\kappa \, \partial T / \partial n + AI = 0 \quad \text{when} \quad z = Z(x, y, t)$$

where $\partial/\partial n$ is the normal derivative. Projecting this onto the plane $z = 0$, differentiating with respect to time, and taking into account the fact, that $\partial Z / \partial t = v_z$, we obtain

$$\frac{\partial^2 T_1}{\partial t \partial z} + v_z \left(\frac{d^2 T_0}{dz^2} \right) = 0 \quad \text{when} \quad z = 0$$

However, according to our assumption (see Equation (7.1)), $d^2 T_0 / dz^2 = 0$, from which it follows that

$$\partial T_1 / \partial z = 0 \quad \text{when} \quad z = 0 \qquad (7.5)$$

An arbitrary smooth perturbation to the shape of the liquid surface and its temperature may be expanded in the Fourier integral. Therefore, when using the principle of superposition in the linear approximation, to determine the evolution of an arbitrary perturbation, it is sufficient to investigate the evolution of the monochromatic perturbation of the form:

$$\begin{aligned}\varphi &= \varphi(z) \exp(ikx + \gamma t) \qquad A_y = A(z) \exp(ikx + \gamma t) \\ T_1 &= T_1(z) \exp(ikx + \gamma t) \end{aligned} \qquad (7.6)$$

Substituting (7.6) into (7.4), we obtain:

$$\left(\frac{d^2}{dz^2} - k^2\right)\varphi(z) = 0$$

$$\left(\frac{d^2}{dz^2} - l_1^2\right)A(z) = 0 \quad (7.7)$$

$$\left(\frac{d^2}{dz^2} - l_2^2\right)T_1(z) = \chi^{-1}\frac{dT_0}{dz}\left(\frac{\partial\varphi(z)}{\partial z} + ikA(z)\right)$$

$$l_1^2 = k^2 + \frac{\gamma}{\nu} \qquad l_2^2 = k^2 + \frac{\gamma}{\chi} \quad (7.8)$$

From the condition that Equation (7.7) should have a solution satisfying the boundary conditions (7.5) and (7.6), the dispersion equation $\gamma(k)$ follows for the perturbations considered. However, the investigation of Equation (7.6) is so difficult for a general case that we will confine our analysis to the case of a "deep liquid" $kH \gg 1$. A more general study is performed in the original work.[4] The dispersion equation, when $kH \gg 1$, becomes

$$\frac{d\sigma}{dT}\frac{dT_0}{dz}(\eta\chi)^{-1}[M(\gamma,l_1) - N(\gamma,k)M(\gamma,k)] + 2N(\gamma,k) = 1 + \left(\frac{l_1}{k}\right)^2 \quad (7.9)$$

where

$$\omega_0^2 = gk + (\sigma/\rho)k^3 \qquad M(\gamma,s) = \chi/\gamma - 1/l_2(l_2 + s)$$

$$N(\gamma,k) = (\omega_0^2 + 2\gamma\nu k l_1)/(\omega_0^2 + \gamma^2 + 2\gamma\nu k^2)$$

and l_1 and l_2 are defined by Equation (7.8). From Equation (7.9) in the limiting case when $(\chi k^2, \nu k^2) \ll (|\gamma|, \omega_0)$ we easily obtain

$$(\gamma^2 + \omega_0^2)(\gamma^2 + c^2 k^2) = 0 \quad (7.10)$$

where

$$c^2 = \left|\frac{d\sigma}{dT}\frac{dT_0}{dz}\right|\rho^{-1}(1 + P^{1/2})^{-1}$$

and $P = \nu/\chi$ is the Prandtl number. From Equation (7.10), it can be seen that the dispersion equation has two independent oscillatory branches in the long-wave limit: gravitational-capillary, with $\gamma_1^{(0)} = i\omega_0(k)$ and thermocapillary with $\gamma_2^{(0)} = ick$. Assuming further that $\gamma_1(k) = i\omega_0(k) + \delta_1(k)$, $\gamma_2(k) = ick + \delta_2(k)$, the relation between modes leading to parametric instability may be obtained. Instability will begin if $\text{Re}\,\delta_{1,2} > 0$. The calculations show that the sign of $\text{Re}\,\gamma_2$ is the same as that of the difference $ck - \omega_0$. As is clear from Equation (7.1), the magnitude of c is proportional to $(AI)^{1/2}$, and ω_0 is independent of the radiation intensity. This means that a certain critical level of intensity I^* exists, above which $\text{Re}\,\delta_2$ becomes positive, i.e., thermocapillary waves are excited. From Figure 7.2 it is clear that the wave numbers of the excited modes lie in the finite range of $k_1(I) < k < k_2(I)$.

CHAPTER 7. INSTABILITIES IN LIQUIDS

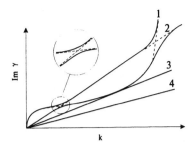

Figure 7.2: Dispersion curves $\operatorname{Im}\gamma(k)$. 1—Gravitational-capillary waves; 2—Thermocapillary waves, $I > I^*$; 3—$I = I^*$; and 4—$I < I^*$.

The behavior of $\operatorname{Re}\delta_1$ is somewhat more complex: it is negative at $\omega_0 < ck$, but when $\omega_0 > ck$, for every k there is a certain minimum $I_m(k)$ above which $\operatorname{Re}\delta_1 > 0$. Calculations show that:

$$\min_k[I_m(k)] = I_m(k_c) = 2^{5/2}\eta k(\omega_c^3/\chi)^{1/2}(k_c A(d\sigma/dT))^{-1}$$

where

$$k_c = (\rho g/5\sigma)^{1/2} \quad \text{and} \quad \omega_c = \omega_0(k_c)$$

We now shall consider the case of resonant interaction of natural modes. This case corresponds to those values of k which are close to the solutions of the equation $\omega_0(k) = kc$. Near resonance the real part of γ increases and ceases to be a small correction to the imaginary part. Calculations show that, for this case,

$$\operatorname{Re}\gamma \simeq (\omega_0/2)(\chi k^2/\omega_0)^{1/4}(1 + P^{1/2})^{1/2}\sin(\pi/8)$$

Whereas, far from resonance, it is possible to distinguish potential (gravitational capillary) and vortex (thermocapillary) modes, such a distinction becomes impossible in the resonance regions. In the developing complex motion, both components—potential and vortex—have the same order of magnitude.

The analysis, performed by Levchenko and Chernyakov[4] for a liquid of finite depth, shows that the qualitative picture of the occurrence of instability described above is preserved completely in the general case. The quantitative difference is in the increase in the threshold intensity I^*, which is determined by the additional dissipation of energy near the bottom boundary of the liquid film.

We will briefly discuss the nonlinear stage of the instability development. From the discussion above, it is clear that the liquid remains motionless, and the temperature profile $T_0(z)$ is linear for radiation intensities $I < I_{min} = I_m(k_c)$. For a small excursion above the threshold: $I > I_{min}$, $I - I_{min} \ll I_{min}$, motion whose characteristic scale is $\lambda_c = 2\pi/k_c$ occurs in the liquid film. It is natural to assume that the growth of the unstable mode in the weakly supercritical case will stabilize due to nonlinear effects; this leads to the formation of a new stationary state with broken symmetry. The spatial structure and the temperature and velocity fields in this state may be different and, for given k_c, there may be rolls, as well as triangular, rectangular, or hexagonal cells.

Finally, it is possible, generally speaking, to have a case where stationary structures are not formed at all, and the motion of the liquid has a complex, chaotic, turbulent form. Of all the possibilities enumerated by Levchenko and Chernyakov[4] only the simplest two-dimensional structure was studied in the form of rolls. If the supercriticality is small, we can assume that an almost monochromatic gravitational-capillary wave occurs whose amplitude Z_k satisfies the equation:

$$\frac{d}{dt}|Z_k|^2 = 2\gamma |Z_k|^2$$

However, γ itself depends on $|Z_k|^2$ in the nonlinear case and may be presented in the form of the expansion:

$$\gamma = \text{Re}\,\delta_1 + \beta |Z_k|^2 + \ldots$$

To calculate the coefficients of this expansion, standard methods of bifurcation analysis[5] are used: the unknown functions are expanded in a formal series in powers of $\epsilon = kZ_k$, where Z_k is the surface displacement amplitude. In performing such calculations, it is convenient to convert to new independent variables:[6] the potential φ and the stream function ψ. The result obtained[4] is that the perturbed mode is soft, and its amplitude is

$$Z_k \sim [I - I_m(k)]^{1/2}$$

when $I > I_{\min}$. The stability of this structure was not investigated.

Strictly speaking, it follows from the study of weakly supercritical structures that the dependence of the surface reflection on the curvature of the surface must be considered (even for normal incidence of the light). The results[4] were obtained neglecting this dependence. However, there is no reason to expect that taking into account radiation diffraction by the perturbed surface will qualitatively change the results, especially the conclusion concerning the soft character of wave excitation.

To conclude this section, we present numerical estimates, which show the orders of magnitude of the fundamental quantities characterizing the instability being examined. For the melting of iron, the threshold value of the absorbed intensity is $AI_{\min} \simeq 10^3$ W/cm^2. For the intensity considerably higher than the threshold value, $AI = 5 \times 10^5$ W/cm^2, the resonance wave number equals $k \sim 500$ cm^{-1}, and the corresponding value of the instability growth rate equals $\text{Re}\,\gamma \sim 5 \times 10^4$ sec^{-1}.

7.2 Corrugation instability of plane evaporation waves in liquids

Effects of instability which destroy the simplest pattern of evaporation with a plane front are especially important if a liquid layer is formed on the solid body surface. As was shown in Chapter 1, this situation is typical for evaporation of metals. Experimental studies show that even with rather uniform laser intensity profile, a plane evaporation front turns out to be unstable.[7,8] The distortion of a plane evaporation front can be attributed in many cases to the material inhomogeneity, spatial fluctuations of laser radiation intensity, and volume vaporization (boiling).[9,10] These effects can be eliminated to a large extent by a proper choice of experimental conditions. A more general cause of liquid-vapor phase boundary distortion is the intrinsic instability of vaporization front that moves in the direction of temperature gradient.[11,12] In Chapter 5 this instability was studied for a solid-vapor phase transition. In this chapter, we consider the interaction of a spatially uniform laser beam of constant intensity with an absorbing liquid layer, and investigate the stability of surface evaporation of a liquid under these conditions.[13,14] We assume that the gravity force (natural or artificial) is directed normal to the liquid surface. We neglect the flow of liquid along the surface of the solid due to the nonuniformity of the reactive pressure, the boiling of liquid, and the temperature dependence of the parameters of liquid. We take into account, however, the finite thickness of the liquid layer.[14,15]

As in the case of sublimation, it is convenient to choose a coordinate frame in which a plane stationary evaporation front is at rest and lies in the plane $z = 0$, the axis z being directed toward the condensed phase. In this framework, the condensed material moves along the z-axis with the stationary velocity $v_z = -V_s$. As in Chapters 1 and 5, the temperature field in the condensed material is described by the heat conduction equation:

$$c\rho\,(\partial T/\partial t + \mathbf{v}\cdot\nabla T) = \mathrm{div}\,(\kappa\nabla T) + \mu I \tag{7.11}$$

with the boundary condition $T \to 0$ when $z \to \infty$ and the energy balance equation on the evaporation front at $z = Z(x, y, t)$

$$L_{eff} V_n \rho = \kappa(\nabla T)_n \tag{7.12}$$

The evaporation rate is related to the surface temperature:

$$V_n = V_0 \exp(-M L_{eff}/k_B T) \tag{7.13}$$

The effective heat of vaporization depends on the local curvature of the phase boundary as (see Equation (5.3)):

$$L_{eff} = L_v - (\sigma/\rho)(1/R_1 + 1/R_2)$$

where R_1 and R_2 are the principal radii of curvature and σ is the surface energy density.

A number of interesting features of laser evaporation may be attributed to the diffraction of monochromatic laser radiation on the relief formed by the disturbances of the evaporated surface. This effect is important (see references 16 and 17) if the perturbation wavelength is close to the laser radiation wavelength. Far from this resonant region, diffraction effects may be neglected, and the intensity of laser radiation may be described by:

$$I = I_0 \exp\left[\mu(Z - z)\right] \tag{7.14}$$

The normal component of the evaporation front velocity is determined by the vaporization and the hydrodynamic motion of the liquid:

$$v_n = V_n + u_n \qquad (7.15)$$

where u_n is the projection of the mass velocity of liquid onto the normal to the phase boundary. The heat conduction equation (7.11) and boundary conditions (7.12) and (7.13) are quite similar to those used in Chapter 5. In the presence of a melt layer on the surface of the evaporated sample, the boundary conditions on the evaporation front and in the sample depth should be supplemented by the boundary condition on the melting front. In the case of metals, the density and heat capacity of the liquid phase are close to those of the solid phase. In our analysis we shall disregard the jumps in thermophysical properties of the condensed phase at the melting front. We also shall disregard the latent heat of fusion, which is much smaller than the latent heat of vaporization. These simplifying assumptions, normal for laser evaporation studies, lead to some distortion of the stationary temperature profile; but, they actually do not affect the results of stability analysis. Thus, in our formulation of the stability problem, among all the differences in the physical properties of liquid and solid phases only one property, namely the fluidity of liquid phase, which is most important for instability development, is taken into account.

The melt may be considered, with sufficient accuracy, as an ideal incompressible liquid. With the purpose of linear analysis, we can restrict ourselves to the consideration of two-dimensional perturbations dependent on the coordinate z and one of the coordinates in the plane of the unperturbed evaporation front (for certainty, y). The melt flow may be described by a stream function $\Phi(y, z, t)$ that satisfies the equation

$$\frac{\partial \Delta \Phi}{\partial t} - \frac{\partial \Phi}{\partial z} \cdot \frac{\partial \Delta \Phi}{\partial y} + \frac{\partial \Phi}{\partial y} \cdot \frac{\partial \Delta \Phi}{\partial z} = 0 \qquad (7.16)$$
$$u_z = \partial \Phi / \partial y \qquad u_y = -\partial \Phi / \partial z$$

On the evaporation front, the change in pressure is attributed to the effect of surface tension, $p_1 = \sigma(1/R_1 + 1/R_2)$, and to the recoil pressure of the evaporated material, $p_2 = b\rho c_s V_n$, where c_s is the sound speed in vapor near evaporation front and the factor $b \approx 1.67$. The factor b takes into account the hydrodynamic boundary conditions on the evaporation front (Anisimov[18] and Chapter 3). The conditions of mass and momentum flux continuity are the boundary conditions for the hydrodynamics equations on the melting front. The continuity condition for the tangential component of velocity follows directly from these conditions.[11] Emphasize that the condition imposed on the tangential component of velocity is in no way attributed to viscosity and is retained when considering inviscid fluid. If one neglects the difference in densities of the solid and liquid, the condition of mass flux continuity is reduced to the continuity of normal velocity component. In the presence of melting front perturbations, the continuity conditions for both, normal and tangential, components of the velocity can not be satisfied in assumption of a potential

CHAPTER 7. INSTABILITIES IN LIQUIDS

flow. Thus, the melt flow turns out to be vortical. The boundary conditions of velocity continuity on the melting surface are equivalent to the conditions of continuity and smoothness of a stream function.

To study the stability of a plane stationary evaporation wave, one should linearize Equations (7.11) through (7.16) in terms of small perturbations of temperature profile, stream function, evaporation rate, and the shape of the evaporation front. By virtue of the superposition principle, it is sufficient to consider disturbances proportional to $\phi_k(z)\exp(iky + \gamma t)$, where k is the perturbation wave number and γ is the instability growth rate. After the linearization, the system of ordinary differential equations for the functions $\phi_k(z)$ is obtained. Solving these equations (with allowance for the boundary conditions on the melting front), we arrive at the system of three equations for the amplitudes of perturbations. It is convenient to write these equations in a matrix form:[14,15]

$$A\varphi = 0 \qquad (7.17)$$

where φ is the column of the amplitudes of evaporation rate perturbations, deformation of the evaporation front, and the mass velocity perturbation, and the matrix A is

$$A = \begin{vmatrix} 1 + m\epsilon^2\chi\omega/V & Z & -\Phi \\ -1 & \gamma & 1 \\ bc_s k & -k(g + \sigma k^2/\rho) & K \end{vmatrix} \qquad (7.18)$$

where $m = cM/k_B$, $\epsilon = k_B T_s(0)/ML_v$, and ω is the root of the equation

$$\omega^2 - V\omega/\chi - (k^2 - \gamma/\chi) = 0 \qquad (7.19)$$

satisfying the condition $\mathrm{Re}\,\omega > 0$;

$$Z = (1 + m\epsilon\chi\omega V)V\sigma k^2/\rho\lambda - \frac{V(\chi\omega - V)(\chi\omega - V - m\epsilon\chi\mu)}{\chi(\omega - \mu) - V}$$

$$K = \frac{(k^2V^2 - \gamma^2)\cosh(kH)}{kV[\cosh(kH) - \exp(-\gamma H/V)] - \gamma\sinh(kH)}$$

$$\Phi = \frac{KV(1 + m\epsilon)\chi\mu P(\mu) - (V + m\epsilon\chi\mu)P(V/\chi)}{2\chi(\chi\mu - V)\cosh(kH)}$$

$$P(p) = \frac{\theta(k,p)}{(kV + \gamma)} + \frac{\theta(-k,p)}{(kV - \gamma)} - \frac{2kV\theta(-\gamma/V,p)}{(k^2V^2 - \gamma^2)}$$

$$\theta(\tau,p) = \exp(\tau H)F(p + \tau + \omega - V/\chi) \qquad F(z) = [1 - \exp(-zH)]/z$$

H is the thickness of the unperturbed melt layer, and $V = V_s$ is the stationary velocity of evaporation front. In the limiting case of significant melt depth,

$kH \gg 1$, for not very quickly damping solutions, $\text{Re}(\gamma) > -kV$, $\text{Re}(\gamma/V + k)H \gg 1$,

$$K \approx kV + \gamma \quad \text{and} \quad \Phi \approx \frac{V(k - \omega - V/\chi - m\epsilon\mu)}{\chi(k+\omega)(k+\omega - V/\chi + \mu)}$$

and, for $H \to 0$

$$K \approx 2V/kH^2 \quad \text{and} \quad \Phi \approx VH/6\chi$$

The conditions for the existence of a nontrivial solution of Equation (7.17),

$$\det(A) = 0, \tag{7.20}$$

is the dispersion equation for the instability growth rate $\gamma(k)$.

Before studying the dispersion equation, it is useful to consider the structure of the matrix (7.18). The upper left minor of the determinant (7.20) corresponds to the problem of pure evaporation (without the melt flow) and describes the sublimation of solids considered in Anisimov et al.[11] and in Chapter 5. Setting this minor equal to zero we obtain the equation:

$$\left[1 + m\epsilon^2(\chi\omega/V)\right]\gamma + Z = 0 \tag{7.21}$$

which, in combination with Equation (7.19) forms a set of equations equivalent to that obtained in Chapter 5.

The bottom right minor of the determinant (7.20) describes the evolution of surface waves in a melt layer at a constant rate of melt evaporation. Stationary liquid flow from the melting front to the vaporization front leads to significant differences of this case from ordinary gravitational-capillary waves. Having equated this minor to zero, we obtain the dispersion equation:

$$\gamma K + (g + \sigma k^2/\rho)k = 0 \tag{7.22}$$

In the case of large melt depth it is reduced to a quadratic equation:

$$\gamma^2 + kV\gamma + (g + \sigma k^2/\rho)k = 0$$

whose solution is

$$\gamma = -kV/2 \pm \sqrt{[(kV/2)^2 - (g + \sigma k^2/\rho)k]}$$

The mass flux to the evaporation front from the melt depth results in the attenuation of gravitational-capillary waves. When $g + \sigma k^2/\rho < 0$, the aperiodic growth of perturbations occurs. When $kV^2/4 > g + \sigma k^2/\rho > 0$, a new mode of the evolution of perturbations—aperiodic attenuation—appears. With the decrease of kH, the oscillatory nonpotential modes become important. A typical form of the dispersion relation for this case is shown in Figure 7.3. The above specific features of the dispersion dependence (7.22) remain also with the complete dispersion equation (7.20).

It is convenient to write the dispersion equation (7.20) in the form:

$$(\gamma + W)K + \left(-VW + g_{eff} + g + \sigma k^2/\rho\right)k = 0 \tag{7.23}$$

where

$$g_{eff} = (bc_s + V)W \quad \text{and} \quad W = (Z + \gamma\Phi)/(1 + m\epsilon^2\chi\omega/V - \Phi) \tag{7.24}$$

CHAPTER 7. INSTABILITIES IN LIQUIDS

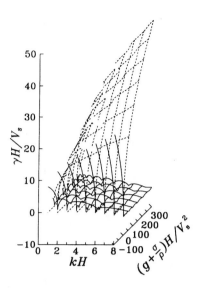

Figure 7.3: The dependence of the instability growth rate γ on the perturbation wave number k and effective gravity acceleration $(g + \frac{\sigma}{\rho}k^2)$ for liquid film evaporation with constant evaporation rate. Solid lines—$\operatorname{Re}\gamma$; dashed lines—$\operatorname{Im}\gamma$.

It follows from (7.24) that, at $k \gg V/\chi$, W turns out to be close to $-\gamma_0$, where γ_0 is the solution of Equation (7.21) corresponding to the case of pure sublimation:

$$\gamma_0 = -Z/(1 + m\epsilon^2 \chi \omega/V).$$

The analysis of Equation (7.23) shows that W depends mainly on k. With the variation of other parameters (g, H) there may occur strong changes in γ and relative variation of W is usually small. We may assume then approximately that $W \approx -\gamma_0$. Since the sound velocity c_s is much larger than the vaporization front velocity V, the main effect of thermal perturbations at large k and H is described by g_{eff}. The sublimation growth rate thus describes the effective mass force affecting the melt. When there is an instability in the sublimation problem ($\gamma_0 > 0$) then aperiodic instability may occur in the complete problem (7.23). The characteristic value of growth rate is greater than γ_0:

$$\gamma \approx (bc_s k \gamma_0)^{1/2}$$

The instability region is determined by the inequality:

$$-g_{eff} \approx bc_s \gamma_0 > g + \sigma k^2/\rho$$

instead of $\gamma_0 > 0$, and becomes, therefore, more narrow than in the case of sublimation. This mode of instability is suppressed completely at any radiation intensity when the (artificial) gravity acceleration is rather high,

$$g > g_c \approx \max\left((V + bc_s)k\gamma_0 - \sigma k^2/\rho\right)$$

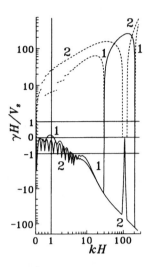

Figure 7.4: Dispersion curves $\gamma(k)$ at high radiation intensity $I_0 = 5 \times 10^7$ W/cm^2 and different values of gravity acceleration: 1—$g = 0$; 2—$g = 790.4 V_s^2/H = 3.232 \times 10^{11}$ cm/s^2. Solid lines— Re γ; dashed lines— Im γ.

To illustrate this, in Figure 7.4, the dispersion curves are given in the cases of absence of gravity force ($g = 0$) and of large enough gravity force ($g > g_c$) when the aperiodic instability is suppressed. Within a wide range of evaporation front velocity V the value of stabilizing acceleration g_c is nearly proportional to V^2 and, therefore, to I_0^2, where I_0 is the laser intensity.

When $k < m\epsilon\mu$, the phase shift between the thermal and hydrodynamic perturbations (which manifests itself in the presence of the imaginary part of γ near the instability threshold) leads to the buildup of surface waves. At small evaporation rate, $V \ll m\epsilon\mu\chi$, this buildup may become stronger than the convective stabilization and evoke the instability. A characteristic form of the dispersion relation in the case of the coexistence of this and the aperiodic instability considered above is shown in Figure 7.5. Hydrodynamic oscillatory instability becomes important at relatively low radiation intensities when the short wavelength aperiodic instability cannot develop. At rather high gravity acceleration, the phase relation between thermal and hydrodynamic disturbances is violated and instability is absent.

Quantitative estimates of the growth rate and stabilizing gravity acceleration for hydrodynamic instability depend on the model used. In particular, a noticeable stabilization may be related to the melt viscosity that was disregarded in the previous analysis. This instability does not arise at a small film thickness either. It should be noted that a film thickness at which hydrodynamic perturbations are stabilized is much larger than the radiation penetration depth or the thickness at which the melt flow becomes negligible, and the dispersion equation reduces to (7.21). In the present consideration, the description of the melt flow is simplified. We do not consider the melt

CHAPTER 7. INSTABILITIES IN LIQUIDS

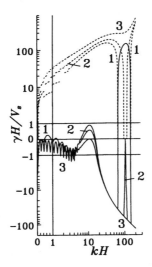

Figure 7.5: Dispersion curves $\gamma(k)$ at $I_0 = 2 \times 10^7$ W/cm^2 and different values of gravity acceleration: 1—$g = 0$; 2—$g = 352.6 V_s^2/H = 1.187 \times 11^{11}$ cm/s^2; 3—$g = 1229 V_s^2/H = 4.139 \times 11^{11}$ cm/s^2. Solid lines—Re γ; dashed lines—Im γ.

flow parallel to the melting front due to the recoil pressure gradient, and we consider the liquid layer thickness, H, as an independent parameter. In this situation, the description of hydrodynamic mode stabilization at a small melt thickness and the transition $\gamma(H) \to \gamma_0$ when $H \to 0$ has a qualitative character.

Note that a finite thickness in the liquid film results in the reduction of the considered hydrodynamic instability and in the appearance of the new instability. The latter is connected to the nonpotential character of melt flow. The growth rate of this instability reaches its maximum at $kH \approx 1$. The dispersion equation for this mode is similar to (7.22); however, the interaction of thermal and hydrodynamic perturbations leads to the growth of perturbation. The growth rate of this instability is usually small, as compared with the growth rates of the previous modes of instability. The acceleration required to stabilize this instability is also relatively small.

As shown by calculations, a plane stationary evaporation front may be stabilized by a sufficiently strong gravity acceleration at arbitrary laser intensity. The dependence of the acceleration necessary to stabilize all of the instability modes considered above on the laser intensity is shown in Figure 7.6. It should be noted that the relative position of the curves corresponding to different instability modes depends on the parameters of material, such as the surface tension, σ, and the absorption coefficient, μ.

In typical cases of the coexistence of the short wavelength aperiodic and oscillatory hydrodynamic instabilities, the latter has a much smaller growth rate; but, to stabilize it, one requires a much higher gravity acceleration. A

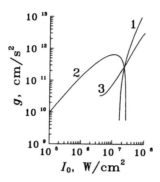

Figure 7.6: The stabilizing acceleration as a function of laser intensity. 1—stabilization of short wavelength aperiodic (sublimation-type) instability; 2—stabilization of oscillating hydrodynamic instability; 3—stabilization of long wavelength nonpotential hydrodynamic modes.

stable evaporation regime can be reached in this case using a short interaction time (not sufficient for the development of relatively slow oscillatory instability) and the artificial gravity acceleration (to stabilize the aperiodic short wavelength instability). Note, however, that the case where the instability develops with short-pulse evaporation requires special study.

REFERENCES

1. Landau, L. D. and Lifshitz, E. M., *Fluid Mechanics*, Pergamon Press, Oxford, 1987

2. Landau, L. D. and Lifshitz, E. M., *Statistical Physics, Part I*, Pergamon Press, Oxford, 1980

3. Pearson, J. R. A., On convection cells induced by surface tension, *Journ. of Fluid Mech.*, **4**, 489, 1959

4. Levchenko, E. B. and Chernyakov, A. L., Instability of surface waves in a nonuniformly heated liquid, *Sov. Phys.–JETP*, **54**, 102, 1981

5. Ioos, G. and Joseph, D. D., *Elementary Stability and Bifurcation Theory*, Springer-Verlag, New York, 1980

6. Stretinskii, L. N., *Theory of Wave Motions of a Liquid*, Moscow, Nauka, 1977 (Russian)

7. Panteleev, V. V. and Yankovskii, A. A., Use of laser beams for vaporizing materials in spectral analysis, *Journ. Appl. Spectroscopy*, **3**, 260, 1965

CHAPTER 7. INSTABILITIES IN LIQUIDS

8. Bass, M., Nassar, M. A., and Swimm, R. T., Impulse coupling to aluminum resulting from Nd:glass laser irradiation induced material removal, *Journ. Appl. Phys.*, **61**, 1137, 1987

9. Rykalin, N. N. and Uglov, A. A., Bulk vapor production by a laser beam acting on a metal, *High Temperature*, **9**, 522, 1971

10. Burmistrov, A. V., Role of bubble boiling in the interaction of intence radiation with matter, *Journ. Appl. Mech. Tech. Phys.*, **20**, 287, 1979

11. Anisimov, S. I., Tribelskii, M. I., and Epelbaum, Ya. G., Instability of a plane evaporation front in the interaction of laser radiation with a medium, *Sov. Phys.-JETP*, **51**, 802, 1980

12. Langer, J. S., Instabilities and pattern formation in crystal growth, *Rev. Mod. Phys.*, **52**, 1, 1980

13. Levchenko, E. B. and Chernyakov, A. L., Stability of plane wave front of fluid evaporation, *Journ. Appl. Mech. Tech. Phys.*, **23**, 870, 1982

14. Khokhlov, V. A., Instability of evaporation of a layer of melt in a gravitational field, *Sov. Phys.-Tech. Phys.*, **33**, 864, 1988

15. Khokhlov, V. A., Instability of laser-induced evaporation of the melt layer in the gravity field, *Stability and Applied Analysis of Continuous Media*, **2**, 547, 1992

16. Akhmanov, S. A., Emel'yanov, V. I., Koroteev, N. I., and Seminogov, V. N., Interaction of powerful laser radiation with the surfaces of semiconductors and metals: nonlinear optical effects and nonlinear optical diagnostics, *Sov. Phys.-Uspekhi*, **28**, 1084, 1985

17. Bonch-Bruevich, A. M., Libenson, M. N., Makin, V. S., and Trubaev, V. V., Surface electromagnetic waves in optics, *Optical Engineering*, **31**, 718, 1992

18. Anisimov, S. I., Vaporization of metal absorbing laser radiation, *Sov. Phys.-JETP*, **27**, 182, 1968

Chapter 8

INSTABILITIES CONNECTED WITH LASER-INDUCED CHEMICAL REACTIONS

In the previous chapters we considered several examples of instabilities appearing in macroscopic laser-matter interactions. In typical cases the mechanism of instability was connected with a positive feedback between the laser-induced temperature rise and the laser energy release. In its simplest form, this feedback can be represented by an algebraic relation of the type of Equation (2.14). According to this equation, the absorption coefficient increases with the temperature increasing. This gives rise to the instability similar to the thermal explosion.[1] As we have seen while discussing the polymer ablation a feedback may have the form of differential equation (see Equation (2.20)). Further examples of the feedback of this type can be found in laser thermochemistry.[2] In this chapter we shall consider two instability problems related to the surface oxidation of metals: (1) the instability of oxide film growth on the surface of uniformly irradiated target, and (2) the laser initiation of surface combustion waves.

8.1 Instability of laser-stimulated surface oxidation of metals. Spatially uniform temperature fields

In this section we shall consider the heterogeneous oxidation reaction of metals in air under the action of CO_2-laser radiation. We shall examine, first, the initial stage of heating of a metallic target, when the oxide layer is thin and one can neglect the interference phenomena leading to oscillations in the surface reflectivity. We shall assume for simplicity that the target is thermally thin, i.e., that the characteristic time of temperature changes is much longer than H^2/χ, where H is the target thickness and χ is the thermal diffusivity. Under this constraint, we may substitute the heat conduction equation with the equation of energy balance in the form:

$$mc\,dT/dt = PA(h) - P_l(T) \qquad P_l = \beta s(T - T_0) \qquad (8.1)$$

where h is the thickness of the oxide layer, c is the specific heat of the oxide (per unit mass), P is the laser power, P_l is the power of convective energy losses, $A(h)$ is the absorptivity of the layered system "metal + oxide film", m is the target mass, and s is the target area. We assume that the oxide film growth obeys the parabolic oxidation law:[3]

$$dh/dt = (d/h)\exp(-T_d/T) \qquad (8.2)$$

where d and T_d are constants. If h is small in comparison with the laser wavelength, the absorptivity $A(h)$ can be written in the form (see Equations (8.8) and (8.9)):

$$A(h) = A_0 + bh^2, \qquad (8.3)$$

where A_0 is the absorptivity of a metal.[4] The initial conditions for Equations (8.1) and (8.2) read

$$T(0) = 0 \quad \text{and} \quad h(0) = 0 \qquad (8.4)$$

We introduce the dimensionless variables

$$y = T/T_d \qquad y_0 = T_0/T_d \qquad \tau = \beta st/mc$$
$$\mu = 2bdPmc/T_d(\beta s)^2 \qquad y_1 = y_0 + PA_0/\beta sT_d$$

Under typical conditions of laser heating, the small parameters of the problem are $y_0 \ll 1$, $y_1 \ll 1$, $\mu \ll 1$. Equations (8.1) and (8.2) can be reduced to a single differential equation

$$d^2y/d\tau^2 + dy/d\tau - \mu\exp(-1/y) = 0 \qquad (8.5)$$

The initial conditions (8.4) are transformed into

$$y(0) = y_0, \qquad dy/dt|_{t=0} = y_1 - y_0 \qquad (8.6)$$

The characteristic feature of the solutions of Equation (8.5) is an avalanche-like acceleration of the oxidation reaction after some induction period, τ_i. This instability is similar to the inflammation process in combustion physics. The induction period usually is defined as the time at which the reaction rate reaches its maximum: $d^2h/dt^2|_{t=\tau_i} = 0$. This time was calculated by Bunkin et al.[2] using the approximate solution of Equations (8.5) and (8.6) by the method of iteration. The result can be represented in the following form:

$$\begin{aligned}\tau_i &\approx (y_1 - y_0)^{-1}\{1/\log[\mu/(y_1 - y_0)] - y_0\} &\text{for } g(\tau_i) \gg 1 \\ \tau_i &\approx (y_1^2/\mu)\exp(-1/y_1) &\text{for } g(\tau_i) \ll 1\end{aligned} \qquad (8.7)$$

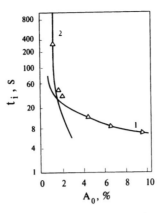

Figure 8.1: Theoretical and experimental dependence of the induction time t_i on metal absorptivity A_0 for oxidation of copper in air by continuous CO_2-laser. 1—the dependence calculated using Equation (8.7), $g \gg 1$ (fast activation); 2—the same for $g \ll 1$ (quasi-stationary mode); the triangles—experimental results[5].

where $g(\tau) = 1/y(\tau) - 1/y_1$. Thus, there are two modes of the oxidation reaction: fast activation, ($g \gg 1$), and quasi-stationary activation ($g \ll 1$). Figure 8.1 shows the comparison of experimental and theoretical dependences of the induction period on the initial absorptivity of the copper target. Experimental data were taken from Arzuov et al.,[5] and the theoretical curves were calculated using Equation (8.7).

A more general kinetic equation for the surface oxidation ($dh/dt \propto h^{-\alpha}$) was considered by Libenson.[6] The induction period there was calculated for the case of quasi-stationary activation.

If the thickness of oxide film is comparable with the laser wavelength, interference variations in the absorptivity of the metal-oxide system can be observed. This is evident from an analysis of the equations for the absorptivity of a layered system[4]

$$\left.\begin{aligned} A(h) = 1 - |r|^2 \quad r &= \frac{r_{12}\exp(-2i\psi) + r_{23}}{\exp(-2i\psi) + r_{12}r_{23}} \quad \psi = 2\pi\sqrt{\epsilon}/\lambda \\ r_{12} = \frac{1-\sqrt{\epsilon}}{1+\sqrt{\epsilon}} \quad r_{13} &= \frac{1-\sqrt{\epsilon_0}}{1+\sqrt{\epsilon_0}} \quad r_{23} = \frac{r_{12} - r_{13}}{r_{12}r_{13} - 1} \\ A_0 = 1 - |r_{13}|^2 \quad &\sqrt{\epsilon} = n - i\kappa \quad \sqrt{\epsilon_0} = n_0 - i\kappa \end{aligned}\right\} \quad (8.8)$$

where r_{12} and r_{13} are the amplitude coefficients of reflection from the oxide and metal, respectively; ϵ and ϵ_0 are the dielectric permittivities of the oxide and metal, respectively; A_0 is the absorptivity of the metal ($A_0 \ll 1$); h is the thickness of the oxide layer; and λ is the laser wavelength. For thin oxide layers, $A(h)$ can be expressed as a series in powers of (h/λ):

$$A(h) = A_0 + a(h/\lambda) + b(h/\lambda)^2 + \ldots \tag{8.9}$$

where the coefficients a, b, etc. are determined by Equations (8.8). In the far infrared region we obtain

$$a \approx \pi A_0^2(n^2 - 1) \quad \text{and} \quad b \approx 4\pi A_0(n^2 - 1). \tag{8.10}$$

Since $A_0 \ll 1$, it follows from (8.10) that $a \ll b$. This means that the change in the absorptivity is determined by the quadratic term in the expansion of $A(h)$, as assumed in (8.3).

For weakly absorbing oxides ($\mu\lambda \ll 1$, $\mu = 4\pi\kappa/\lambda$) a simple approximate expression can be derived[5] from (8.8) for $A(h)$

$$A(h) \approx \frac{n^2 A_0 + 2\kappa[ph - \sin(ph)]}{n^2 + (1 - n^2)\sin^2(ph/2)} \qquad p = 4\pi n/\lambda.$$

For the total number of interference oscillations, which can be observed in laser oxidation experiments, one can obtain a simple estimate (from the condition $\mu h \approx 1$): $N \approx n/2\pi\kappa$. For example, for Cu_2O at the wavelength $\lambda = 10.6\mu m$, $n = 2.45$, $\kappa = 0.027$, we have $N \approx 14$. A more complicated situation was analyzed by Bunkin et al.,[7] where the two-layer system ($Cu_2O + CuO$) was considered.

8.2 Instability of surface oxidation of metals: Two-dimensional effects

In this section we shall show that the above considered one-dimensional growth of the oxide layer is unstable with respect to two-dimensional perturbations. We shall consider a thermally thin metallic plate and describe the temperature field by the heat conduction equation:

$$\partial T/\partial t = \chi \Delta T + A(h)I/Hc\rho \tag{8.11}$$

where χ is the thermal diffusivity of metal, $c\rho$ is the heat capacity (per unit volume), Δ is the Laplace operator in two dimensions (x,y), and $A(h)$ is the absorptivity of the layered system (metal + oxide). For simplicity, we will disregard the heat losses and assume the simplest linear oxidation law:

$$dh/dt = d\exp(-T_d/T) \tag{8.12}$$

where d and T_d are constants. If the laser intensity, I, is constant, Equations (8.11) and (8.12) have a particular solution $T = T_s(t)$, $h = h_s(t)$. We will study the stability of this solution with respect to small perturbations

$$\begin{aligned} \delta T(y,t) &= T_1 \exp\left[iky + \int_0^t \gamma(\tau)d\tau\right] \\ \delta h(y,t) &= h_1 \exp\left[iky + \int_0^t \gamma(\tau)d\tau\right] \end{aligned} \tag{8.13}$$

Here, $\gamma(t)$ is a slow-varying function of time. A similar form of perturbations was used in Section 5.2, where the corrugation instability of nonstationary

CHAPTER 8. INSTABILITIES WITH CHEMICAL REACTIONS

metal vaporization was studied. Calculations similar to those performed in Section 5.2 result in the following dispersion equation (see also references 8 and 9).

$$\gamma(k) = -\chi k^2/2 + \left[(\chi k^2/2)^2 + (T_d\chi I/\kappa H)\dot{h}\, dA/dh\right]^{1/2} \quad (8.14)$$

where $\dot{h} = dh/dt$. Due to the interference effects discussed in previous section, the derivative dA/dh is an oscillating function of h. One can see from (8.14) that the instability growth rate $\gamma(k)$ is positive for a positive dA/dh term. The maximum of $\gamma(k)$ is reached at $k = 0$, i.e., the thermal conduction produces a stabilizing effect. As the oxide film thickness, h, increases, the derivative dA/dh changes its sign. When dA/dh is negative, the real part of the instability growth rate, $\mathrm{Re}\,\gamma(k)$, also becomes negative. This means that the perturbations, which were growing during the stage $dA/dh > 0$, begin to decay exponentially. We see, thus, that the stages of stable and unstable perturbation growth should change periodically. The evolution of monochromatic initial perturbation is determined (in adiabatic approximations) by the value of the integral:

$$S(k,\tau) = \int_0^\tau \mathrm{Re}\,\gamma(k,t)\, dt, \quad (8.15)$$

where τ is the laser pulse duration. The irreversible nonlinear stage of perturbation growth is reached during the laser pulse, if the integral (8.15) is of the order of unity, $S(k,\tau) \geq 1$. Consider the dispersion equation (8.14) in the case of low laser intensity, when

$$\left|(T_d\chi I/\kappa HT^2)\dot{h}\, dA/dh\right| \ll (\chi k^2/2)^2 \quad (8.16)$$

Expanding $\gamma(k)$ in series in terms of small I and confining ourselves to the linear term, we have:

$$\gamma(k) \approx (T_d I/\kappa HT^2 k^2)\dot{h}\, dA/dh \quad (8.17)$$

Substituting (8.17) into (8.15), we obtain

$$S(k,\tau) \approx AT_d I/\kappa HT^2 k^2 \ll 1 \quad (8.18)$$

The latter inequality can be proved in the following way. It follows from (8.12) and (8.11) that

$$\dot{h}/h \approx \ddot{h}/\dot{h} \approx T_d\dot{T}/T^2 \approx T_d\chi AI/\kappa HT^2.$$

Using this equation, the left side of Equation (8.16) can be rewritten in the form:

$$\left|(T_d\chi I/\kappa HT^2)\dot{h}\, dA/dh\right| \approx (2nh/\lambda)\langle A^2\rangle(T_d\chi I/\kappa HT^2)^2 \quad (8.19)$$

where n is the real part of the refraction index of oxide film; λ is the wavelength of laser radiation; and $\langle A^2 \rangle$ is the average value of A^2 (over several interference

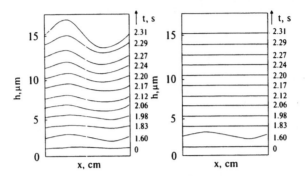

Figure 8.2: The evolution of the oxide film on a thermally thin copper target. The values of time are indicated on the right. Normal incidence of CO_2-laser radiation. a—supercritical mode, b—subcritical mode.

oscillations). The inequality (8.18) follows immediately from (8.16) and (8.19). We see, then, that at low laser intensities and arbitrary laser pulse lengths, the growth of the oxide film is stable, despite the fact that there are the stages when the perturbations (8.13) increase with time.

We now will consider the case of high laser intensities, when

$$\left|(T_d\chi I/\kappa HT^2)\dot{h}\, dA/dh\right| \geq (\chi k^2/2)^2 \tag{8.20}$$

If $dA/dh > 0$, the growth rate $\gamma(k)$ is real and increases as $I^{1/2}$ with increasing I. If $dA/dh < 0$ and the inequality (8.20) is fulfilled, $\gamma(k)$ is complex, and its real part, $\text{Re}\,\gamma(k) = -\chi k^2/2$, does not depend on I. It is clear that if the laser intensity exceeds some critical value I^*, the amplification of perturbations during the stages with $dA/dh > 0$ should be greater than the damping of perturbations during the stages with $dA/dh < 0$. This means that the growth of the oxide layer is unstable. A rough estimate for I^* follows from (8.20) and (8.19):[8]

$$I^* \approx (\lambda/2nh)^{1/2}\kappa HT^2 k^2/2\langle A\rangle T_d \tag{8.21}$$

A more precise determination of I^* can be made using numerical calculations. Such calculations were performed by Gol'berg.[9] The oxidation kinetics were described by the parabolic oxidation law (8.2). This results in the dispersion equation

$$\gamma(k) = -(\chi k^2 + \dot{h}/h)/2$$
$$+ \left[(\chi k^2 + \dot{h}/h)^2/4 + (T_d I\chi/\kappa HT^2)\dot{h}\, dA/dh - \chi k^2 \dot{h}/h\right]^{1/2}$$

which differs only slightly from (8.14). All the previously discussed qualitative results obtained using the linear oxidation law (8.12) remain valid for the parabolic law (8.2). In Figure 8.2 the nonlinear evolution of the shape of oxide film is shown. The pictures were obtained as the result of numerical solution of Equations (8.11) and (8.2) for a thermally thin copper target. The initial conditions and parameters were chosen as:

$$h(0,x) = h_0 + \delta h \qquad h_0 = 1 \times 10^{-5} \text{ cm} \qquad \delta h = 0.25 h_0 \sin kx$$
$$T(0,x) = 300 \text{ K} \qquad I/H = 8.2 \text{ kW/cm}^3$$
$$\text{case } (a) \quad k = 1.3 \text{ cm}^{-1} \qquad \text{and} \qquad \text{case } (b) \quad k = 5.3 \text{ cm}^{-1}$$

Since the critical value of laser intensity depends on the perturbation wavelength (see (8.21)), the same intensity I appears to be supercritical in case (a) and subcritical in case (b). Correspondingly, we see the rapid growth of the initial perturbation in the case (a) and the damping of the initial perturbation in the case (b).

8.3 Laser-initiated surface combustion waves

When laser radiation impinges on a target whose dimensions are greater than the focal spot size, heat transport effects play an important role. They can lead to a spatial instability in the form of a propagating reaction front. This situation is characteristic of heterogeneous exothermal reactions in which, under certain conditions, a "hot spot" generated by laser irradiation begins to spontaneously expand. Let a laser beam with radius r_0 be incident on a metallic target occupying the half-space $z > 0$. We assume that an exothermal reaction with linear kinetics (8.12) occurs on the surface of the target. The temperature field in the target is described by the following boundary value problem:[2]

$$\left.\begin{array}{l}\partial T/\partial t = \chi \left[(1/r)\partial/\partial r(r\partial T/\partial r) + \partial^2 T/\partial z^2\right] \\ -\kappa \partial T/\partial z\big|_{z=0} = AI(r) + \rho W d \exp(-T_d/T) \\ T \to 0 \qquad \text{for } r, z \to \infty \end{array}\right\} \qquad (8.22)$$

Here W is the specific heat of the reaction, ρ is the density of the oxide, and d and T_d are the constants in the linear oxidation law. We shall assume that the laser intensity distribution $I(r)$ has the form

$$I(r) = I_m \exp(r^2/r_0^2). \qquad (8.23)$$

The analysis shows that the boundary value problem (8.22) and (8.23) has stationary solutions if the laser intensity I_m does not exceed a certain threshold. The region in the plane (I_m, r_0) where the localized combustion regime is stable can be found from the stationary solution of Equations (8.22). Outside the region of stability, the problem (8.22) does not have stationary solutions.

The stationary surface temperature near the axis of the beam can be written as:

$$T(r,0) \approx T_1(1 - r^2/R^2), \qquad r \ll R \qquad (8.24)$$

The reaction energy release near the axis can be represented in a Gaussian form similar to (8.23):

$$I_c = \rho W d \exp(-T_d/T) \approx \rho W d \exp(-T_d/T_1) \exp(-r^2/b^2) \qquad (8.25)$$

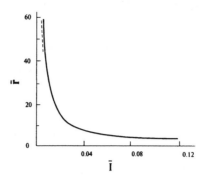

Figure 8.3: The boundary of stability of localized oxidation in the plane $(\bar{r}, \bar{I})$. Above the boundary, a surface combustion wave appears.

where $b^2 = R^2 T_1/T_d$. Since the heat conduction equation (8.22) is linear, its stationary solution can be written in the form:
$$T(r,0) = T_0(r) + T_c(r),$$
where
$$T_0(r) = (\sqrt{\pi}/2) q T_d \exp(-r^2/2r_0^2) I_0(r^2/2r_0^2) \quad (8.26)$$
is the temperature field arising from absorption of laser radiation (8.23), and
$$T_c(r) = (\sqrt{\pi}/2) b \mu T_d \exp(-T_d/T_1) \exp(-r^2/2b^2) I_0(r^2/2b^2) \quad (8.27)$$
is the temperature field arising from the exothermic reaction (8.25). In (8.26) and (8.27), $I_0(z)$ is the modified Bessel function of the zeroth order. We have introduced the parameters
$$q = A I_m r_0 / \kappa T_d \quad \text{and} \quad \mu = \rho W d r_0 / \kappa T_d$$
Expanding (8.26) and (8.27) in powers of r near the beam axis and employing (8.24) and (8.25), we obtain a set of equations for T_1 and R. It can be shown that this system of equations does not have real solutions for all values of q and μ. To find the boundary of the region where stationary solutions of (8.22) exist, we equate the determinant of the set of equations to zero. The result can be written[2] in parametric form as:
$$\begin{cases} q = 0.55(1-\alpha)(0.28+\alpha^2) \\ \mu = 0.30(1-\alpha)(0.28+\alpha^2)^3(0.60-\alpha)\exp\left[\dfrac{3.20(1+\alpha)}{(0.28+\alpha^2)}\right] \end{cases} \quad (8.28)$$
where α is a parameter. Equations (8.27) determine the critical value of the dimensionless laser spot radius, $\bar{r} = r_0(\rho W d/\kappa T_d)$, as a function of the dimensionless laser intensity, $\bar{I} = I_0(A/\rho W d)$. Figure 8.3 shows the boundary of the stability region. Below the stability boundary, the oxidation process is localized spatially and the temperature field is stationary. Above the boundary, the high-temperature region expands with time.

CHAPTER 8. INSTABILITIES WITH CHEMICAL REACTIONS

Note that the appearance of a surface combustion wave has been observed experimentally by Arzuov et al.,[10] where CO_2-laser stimulated combustion of tungsten in air was studied. At supercritical laser intensity ($I_0 \approx 50$ kW/cm^2), a jump-like propagation of the combustion front was observed after the induction period on the order of 1 s.

REFERENCES

1. Frank-Kamenetskii, D. A., *Diffusion and Heat Transfer in Chemical Kinetics*, Plenum Press, New York, 1969

2. Bunkin, F. V., Kirichenko, N. A., and Luk'yanchuk, B. S., Thermochemical action of laser radiation, *Sov. Phys.–Uspekhi*, **25**, 662, 1982

3. Hauffe, K., *Reaktionen in und an festen Stoffen*, Springer Verlag, Berlin, 1955

4. Born, M. and Wolf, E., *Principles of Optics*, Pergamon Press, Oxford, 1970

5. Arzuov, M. I., Barchukov, A. I., Bunkin, F. V., Kirichenko, N. A., Konov, V. I., and Luk'yanchuk, B. S., An Influence of Interference Effects in Oxide Films on Dynamics of Heating of Metals by the Laser Radiation, *Sov. Journ. Quant. Electron.*, **9**, 281, 1979

6. Libenson, M. N., Thermochemical instability in an optically heated condensed medium, *Sov. Tech. Phys. Lett.*, **4**, 370, 1978

7. Bunkin, F. V., Kirichenko, N. A., Konov, V. I., and Luk'yanchuk, B. S., Interference effects in laser heating of metals in an oxidizing medium, *Sov. Journ. Quant. Electron.*, **10**, 891, 1980

8. Tribel'skii, M. I., *Thermal Instabilities in Laser-Matter Interaction*, Doctor of Sci. Dissertation (D.Sci. Thesis), Landau Institute for Theoretical Physics, Chernogolovka, 1984

9. Gol'berg, S. M., *Instability of Vaporization and Surface Oxidation of Solids under Action of Laser Radiation*, Candidate of Sci. Dissertation (Ph.D. Thesis), Landau Institute for Theoretical Physics, Chernogolovka, 1982

10. Arzuov, M. I., Barchukov, A. I., Bunkin, F. V., Konov, V. I., and Lyubin, A. A., Violent surface oxidation of metals and associated phenomena resulting from continuous irradiation with CO_2 laser radiation, *Sov. Journ. Quant. Electron.*, **5**, 931, 1975

CONCLUSION

The theoretical analysis summarized in this book shows that the processes of laser-matter interaction at moderate laser intensities can be unstable. The analysis of instability carried out previously was based on a simplified model description of laser absorption, energy transfer, and phase transitions in laser-heated material. It is important to understand to what extent the theoretical predictions agree with the vast amount of experimental material on laser-induced breakdown of dielectrics and laser ablation.

The main focus in the previous analysis has been on the three different types of instabilities:

(1) the hydrodynamic instability of vapor expansion into a vacuum or ambient gas;

(2) the thermal instability (similar to the "thermal explosion") arising due to the temperature dependence of the absorption coefficient;

(3) the thermal instability of the phase transition front propagating in the direction of the temperature gradient.

We now will discuss some of the experiments which can be interpreted as demonstrating the occurrence of these instabilities.

Consider, first, the instabilities of vapor flow. When the vapor plume expands into an ambient gas, a shock wave is formed in front of the contact surface. In many cases of practical interest, the density of adiabatically expanding vapor is higher than that of the shock-compressed ambient gas. Under this condition, the contact boundary (whose velocity decreases with time) appears to be unstable. This instability, known as the Rayleigh-Taylor instability (RTI), leads to the distortion of the contact surface and turbulent mixing of the vapor and ambient gas. There are many experimental works in which this instability and mixing were observed. We would like to mention only the recent papers[1-4] devoted to excimer laser ablation of different materials (including high-T_c superconductors). Spectroscopic and fast photographic methods were employed in these works. It was shown that at relatively high ambient gas pressure ($p_0 > 1$ mbar) a strong mixing occurs at the contact surface. If the gaseous atmosphere is chemically active, the mixing stimulates the chemical reactions between the vapor and gas. In the case of high-T_c superconductors the conditions are apparently well suited to the formation of diatomic metal oxides which are observed by spectroscopic methods.[3] Similar effects appearing when detonation products expand into air were studied by Kull et al.[5]

Note that the Rayleigh-Taylor instability plays a very important role in processes of laser-driven inertial confinement fusion (ICF).[7] Shell targets used in ICF experiments are susceptible to break up from RTI. The effects of RTI on the compression of ICF targets were studied extensively in the last two decades. However, the effects of RTI and turbulent mixing on the structure and optical properties of laser plumes were discussed mainly on qualitative level. More detailed experimental studies of these effects might be used to check semiempirical theories of turbulent mixing and to determine the values of empirical constants which enter into these theories.

We turn now to the instability of thermal-explosion type. This instability arises if the volume energy release exceeds the energy loss due to heat conduction.[8] Maximum growth rate $\gamma(k)$ is reached at $k = 0$ when the heat conduction does not play role. There are two manifestations of this instability mechanism in laser experiments: (1) laser-induced thermal breakdown of transparent media, and (2) volume and surface thermochemical instability.

A theoretical analysis of the breakdown of transparent media stimulated by small absorbing inclusions is described in Chapter 2. We emphasize that the thermal-explosion behavior is connected with strong nonlinear temperature dependence of the absorption coefficient of dielectric. An inclusion plays the role of a seed, producing local heating of transparent material. Optical breakdown of dielectrics containing small absorbing inclusions was studied in detail experimentally (references 9 and 10). Most of the experiments were conducted with optical glasses. The comparison of theoretical and experimental results conducted by Aleshin et al.[10] shows that calculated breakdown thresholds and induction periods are in reasonable agreement with experimental data. The model of inclusion-stimulated breakdown predicts the dependence of breakdown threshold on the size of the focal volume. This dependence was observed experimentally both for gases and solid materials.[10-13] Since in many cases the characteristics of small absorbing inclusions are not well known, the inverse problem arises: to determine the optical and statistical properties of the ensemble of inclusions by measuring the dependence of the breakdown threshold on the size of focal volume and on the laser pulse length. This problem was studied by Danilenko et al.[14] and Koldunov et al.[15] Note that, from a mathematical point of view, this problem is not well posed. Special mathematical methods have to be applied to solving problems of such type.[16]

Thermochemical instabilities are typical for CO_2-laser interaction with metals, semiconductors, and organic polymers. The laser-induced heating stimulates chemical reactions in the volume or on the surface of target. The rate of chemical reactions usually increases with temperature. If the products of a chemical reaction have a higher absorption coefficient than the initial substance, positive feedback arises, resulting in the instability. A well-known example of volume reaction leading to the thermochemical instability is the formation of soot as the result of thermal decomposition of organic polymers. Theoretical analysis of this instability mechanism was carried out by Tribel'skii and Liberman.[17] This analysis was developed further and experimentally checked by Gol'berg et al.[18] It was shown that the model[17] provides a correct description of the instability development, and the calculated and measured values of the induction period are in good agreement.

Laser-induced reactions of surface oxidation of metals were studied in great detail, both theoretically and experimentally. Interest in the study of these reactions was stimulated by the technological applications of CO_2-lasers, whose radiation is highly absorbed by metal oxides, but is reflected from metal surfaces. Various modes of oxidation and the influence of oxide layers on the dynamics of CO_2-laser heating of metals were examined (references 19–21). Variations in reflectivity during the laser-induced metal oxidation were used to study the kinetics of oxidation.[22] The instability of laser-induced oxidation of metals was first considered by Anisimov et al.[23] The results of numerical simulation of unstable oxide film growth taking into account the interference phenomena were described by Gol'berg and Tribel'skii.[24] The theoretical prediction of the thermochemical instability was confirmed experimentally by Alimov et al.[25] The surface structures in the form of parallel bands whose period ranges from 13 to 200 μm, as well as localized disturbances of film thickness having a diameter about 100 μm were observed. A more detailed discussion of the problem of instability and structures formation, resulting from laser-stimulated heterogeneous chemical reactions, may be found in the literature.[26–28]

Now we will discuss the corrugation instability of a plane phase-transition front. This instability appears when the front of an energy-absorbing phase transition moves in the direction of temperature gradient. This instability is of thermal origin; it is expected to occur in various experiments connected with fast heating of condensed material (e.g., in shock wave compression of solids, fast Joule heating of metals, etc.). Similar instability is well known in the theory of crystal growth.[29] It was shown by Anisimov et al.[30] that plane vaporization front is unstable if the laser intensity exceeds some critical value. When the supercriticality is small, the nonlinear development of this instability leads to the formation of a stationary vaporization wave, moving at constant velocity. At higher intensities, the formation of nonstationary evaporation waves may be expected. Fluctuations of the recoil pressure associated with this vaporization mode lead to the dispersal of target material. As a result, some part of the material is ablated in a condensed state. Many experimental facts are in qualitative agreement with this picture. The experimentally measured specific energy of laser ablation is several times less than the specific heat of vaporization.[31–33] An analysis of the phase composition of the ablation products[32] shows that a portion of the target material is removed in the liquid state. The measurements of target reflectivity (references 34–37) show that the total reflectivity, and especially its specular component,[37] is strongly reduced during the laser pulse. These facts can be explained by the instability of a plane vaporization wave.

The above experiments may be considered as an indirect indication that the corrugation instability plays a role in the laser-matter interaction processes. As shown in Chapter 5, the spatial scale of perturbations having maximum growth rate is of the order of light penetration depth. For metals, it equals about 10^{-5} cm. Direct experimental study of this small-scale instability presents considerable difficulties. Anisimov et al.[38] studied the instability of thin film evaporation both theoretically and experimentally. In this case the spatial scale of growing perturbations is of the order of film thickness. Aluminum foils 10 to 50 μm thick were irradiated by CO_2 pulsed laser. Periodic structures in the form of concentric rings were observed on both the front (irradiated) and rear sides of the target. The structures appear to be independent of a change in laser polarization. Similar structures were observed by Zuev et al.[37] Complicated evaporation modes which may be interpreted as oscillatory and "spin" regimes were reported by Kuznetsov et al.[40] and Duley and Young.[41]

We have not considered structures of another type that were also detected on the surface of metals subjected to laser radiation. These structures resemble a system of parallel bands whose period is on the order of the laser wavelength. The bands were oriented perpendicular to the direction of the electric vector of incident light wave (references 42 and 43). The most plausible explanation of the origin of these structures is the interference between the incident electromagnetic wave and the surface electromagnetic wave generated at the metal-vacuum boundary. The theory of this phenomenon is reported by Bonch-Bruevich et al.[44]

REFERENCES

1. Gilgenbach, R. M. and Ventzek, P. L. G., Dynamics of excimer laser ablated aluminum neutral atom plumes measured by dye laser resonance photography, *Appl. Phys. Lett.*, **58**, 1597, 1991

2. Scott, K., Huntley, J. M., Phillips, W. A., Clarke, J., and Field, J. E., Influence of oxygen pressure on laser ablation of $YBa_2Cu_3O_{7-x}$, *Appl. Phys. Lett.*, **57**, 922, 1990

3. Dyer, P. E., Issa, A., and Key, P. H., Dynamics of excimer laser ablation of superconductors in an oxygen environment, *Appl. Phys. Lett.*, **57**, 186, 1990

4. Geohegan, D. B., Time-resolved diagnostics of excimer laser-generated ablation plasmas used for pulsed laser deposition, *Excimer Lasers*, NATO Inst. for Adv. Study, Elounda, Greece, 1993

5. Kull, A. L., Ferguson, R. E, Chein, K.-Y., Collins, J. P., and Oppenheim, A. K., Gasdynmic model of turbulent combustion in an explosion, *Intern. Conf. on Combustion (Zel'dovich Memorial)*, Moscow, 1994

6. Dyer, P. E. and Sidhu, J., Spectroscopic and fast photographic studies of excimer laser polymer ablation, *Journ. Appl. Phys.*, **64**, 4657, 1988

7. Bodner, S. E., Emery, M. H., and Gardner, J. H., The Rayleigh-Taylor instability in direct-drive laser fusion, *Plasma Phys. and Controlled Fusion*, **29**, 1333, 1987

8. Frank-Kamenetskii, D. A., *Diffusion and Heat Transfer in Chemical Kinetics*, Plenum Press, New York, 1969

9. Manenkov, A. A. and Prokhorov, A. M., Laser-induced damage in solids, *Sov. Phys.-Uspekhi*, **29**, 104, 1986

10. Aleshin, I. V., Anisimov, S. I., Bonch-Bruevich, A. M., Imas, Ya. A., and Komolov, V. L., Optical breakdown of transparent media containing microinhomogeneities, *Sov. Phys.-JETP*, **43**, 631, 1976

11. Cowan, L. H. and Nielsen, P. E., Focal spot size dependence of gas breakdown induced by particulate ionization, *Appl. Phys. Lett.*, **22**, 408, 1973

12. Bettis, J. R., House II, R. A., and Guenther, A. H., Spot size and pulse duration dependence of laser-induced damage, *NBS Special Publ.* 462, 338, 1976

13. Van Stryland, E. W., Soileau, M. J., Smirl, A. L., and Williams, W. W., Pulse-width and focal volume dependence of laser-induced breakdown, *NBS Special Publ.* 620, 375, 1980

14. Danileiko, Yu. K., Minaev, Yu. P., and Sidorin, A. V., Inverse problem of laser breakdown statistics, *Sov. Journ. Quantum Electron.* **14**, 511, 1984

15. Koldunov, M. F., Romanov, M. F., and Filimonov, D. A., Statistical theory of optical breakdown: determination of the density of inclusions from experimental data, *Sov. Phys.-Doklady*, **31**, 638, 1986

16. Tikhonov, A. N. and Arsenin, V. Ya., *Solutions of ill-posed problems*, John Wiley & Sons, Washington, 1984

17. Tribel'skii, M. I. and Liberman, M. A., The role of chemical reactions in the laser destruction of transparent polymers, *Sov. Phys.-JETP*, **47**, 99, 1978

18. Gol'berg, S. M., Matyushin, G. A., Petukhov, A. V., Pilipetskii, N. F., Savanin, S. Yu., and Tribel'skii, M. I., *Study of Thermochemical Instability of Transparent Media Initiated by Absorbing Inclusions*, Preprint $N205$, IPM AN SSSR, Moscow, 1982 (Russian);
    Gol'berg, S. M., Matyushin, G. A., Pilipetsky, N. F., Savanin, S. Yu., Sudarkin, A. N., and Tribelsky, M. I., Thermochemical instability of transparent media induced by an absorbing inclusion, *Appl. Phys.*, **B31**, 85, 1983

19. Volod'kina, V. L., Krylov, K. I., Libenson, M. N., and Prokopenko, V. T., Heating of an oxidizing metal by CO_2-laser radiation, *Sov. Phys.-Doklady*, **18**, 335, 1973

20. Arzuov, M. I., Bunkin, F. V., Kirichenko, N. A., Konov, V. I., and Luk'yanchuk, B. S., *Effect of Surface Oxidation on the Dynamics of Metal Heating by CO_2-Laser Radiation*, Lebedev Physics Institute Preprint $N39$, Moscow, 1978, 37 pp. (Russian)

21. Bonch-Bruevich, A. M., Dorofeev, V. G., Libenson, M. N., Makin, V. S., Pudkov, S. D., and Rubanova, G. M., Exothermic oxidation of metals heated by a light pulse, *Sov. Phys.-Tech. Phys.*, **27**, 686, 1982

22. Alimov, D. T., Bobyrev, V. A., Bunkin, F. V., Zhuravskii, V. L., Luk'yanchuk, B. S., Morozova, E. A., Ubaidullaev, S. A., and Khabibullaev, P. K., Nonequilibrium kinetics of the growth of oxide-layer grains during laser heating of metals in air, *Sov. Phys.-Doklady*, **29**, 1045, 1986

23. Anisimov, S. I., Buzykin, O. G., Burmistrov, A. V., Gol'berg, S. M., and Tribel'skii, M. I., Unstable growth of oxide films induced by laser heating of metals., *Proc. of the All-Union Conference on Nonresonant Laser-Matter Interaction*, Leningrad, 1981, p.290 (Russian)

24. Gol'berg, S. M. and Tribel'skii, M. I., Instability of absorbing oxide film on the surface of metal, *Sov. Tech. Phys. Lett.*, **8**, 178, 1982

25. Alimov, D. T., Atabaev, Sh., Bunkin, F. V., Zhuravskii, V. L., Kirichenko, N. A., Luk'yanchuk, B. S., Omel'chenko, A. I., and Khabibullaev, P. K., Thermochemical instabilities in heterogeneous processes stimulated by laser radiation, *Poverkhnost'*, No 8, 12, 1982 (Russian)

26. Bonch-Bruevich, A. M. and Libenson, M. N., Nonresonant laser chemistry in the interaction of intense radiation with matter, *Bull. Acad. Sci. USSR, Phys. Ser.*, **46**, 82, 1982

27. Bunkin, F. V., Kirichenko, N. A., and Luk'yanchuk, B. S., Thermochemical action of laser radiation, *Sov. Phys.-Uspekhi*, **25**, 662, 1982

28. Bunkin, F. V., Kirichenko, N. A., and Luk'yanchuk, B. S., Structures in laser oxidation of metals, *Sov. Phys.-Uspekhi*, **30**, 434, 1987

29. Langer, J. S., Instabilities and pattern formation in crystal growth, *Rev. Mod. Phys.*, **52**, 1, 1980

30. Anisimov, S. I., Tribelskii, M. I., Epelbaum, Ya. G., Instability of plane evaporation front in interaction of laser radiation with a medium, *Sov. Phys.-JETP*, **51**, 802, 1980

31. Ready, J. F., *Effects of High-Power Laser Radiation*, Academic Press, New York–London, 1971

32. Panteleev, V. V. and Yankovskii, A. A., Use of laser beams for vaporizing materials in spectral analysis, *Journ. Appl. Spectroscopy*, **3**, 260, 1965

33. Anisimov, S. I., Bonch-Bruevich, A. M., Elyashevich, M. A., Imas, Ya. A., Pavlenko, N. A., and Romanov, G. S., Effect of powerful light fluxes on metals, *Sov. Phys.-Tech. Phys.*, 11, 945, 1967

34. Bonch-Bruevich, A. M., Imas, Ya. A., Romanov, G. S., Libenson, M. N., and Mal'tsev, L. N., Effect of a laser pulse on the reflecting power of metals, *Sov. Phys.-Tech. Phys.*, 13, 640, 1968

35. Anisimov, S. I., Imas Ya. A., Romanov, G. S., and Khodyko, Yu. V., *Action of High-Power Radiation on Metals*, Natil. Techn. Inform. Service, Springfield, VA, 1971

36. Zavecz, T. E., Saifi, M. A., and Notos, M., Metal reflectivity under high-intensity optical radiation, *Appl. Phys. Lett.*, 26, 165, 1975

37. Bonch-Bruevich, A. M., Imas, Ya. A., Libenson, M. N., and Shandybina, G. D., Variation in the reflection indicatrices of metals on heating by neodimium-laser radiation pulse, *Bull. Acad. Sci. USSR, Phys. Ser.*, 49, 119, 1985

38. Anisimov, S. I., Gol'berg, S. M., Kulikov, O. L., Pilipetskii, N. F., and Tribel'skii, M. I., New type of laser-evaporation instability, *Sov. Tech. Phys. Lett.*, 9, 99, 1983

39. Zuev, I. V., Selishchev, S. V., and Skobelkin, V. I., Self-oscillations under the action of concentrated energy sources on a material, *Sov. Phys.-Dokl.*, 25, 1021, 1980

40. Kuznetsov, A. E., Orlov, A. A., and Ulyakov, P. I., Pulsating conditions in the evaporation of optical materials under the influence of CO_2 laser radiation, *Sov. Journ. Quantum Electron.*, 2, 44, 1972

41. Duley, W. W. and Young, W. A., Kinetic effects in drilling with the CO_2 laser, *Journ. Appl. Phys.*, 44, 4236, 1973

42. Emmony, D. C., Howson, R. P., and Willis, L. J., Laser mirror damage in germanium at 10.6 μm, *Appl. Phys. Lett.*, 23, 598, 1975

43. Bahin, G., Bernstein, T., and Kalish, R., Laser annealing of indium-implanted $Pb_{0.8}Sn_{0.2}Te$ films, *Appl. Phys. Lett.*, 34, 486, 1979

44. Bonch-Bruevich, A. M., Libenson, M. N., Makin, V. S., and Trubaev, V. V., Surface electromagnetic waves in optics, *Optical Engineering*, 31, 718, 1992

Index

A

ablation 35, 56, 57, 88, 89, 91, 125, 135, 137
   cold 39, 88
   front 36, 88, 89, 91
      perturbed 89
      temperature 38
      velocity 36
   kinetics 91
   of polymers 35, 88
   photochemical 35
   photophysical 36
   rate 39, 88, 91
   threshold 35
   velocity 35–38, 91
absorbing inclusions 25, 26, 28, 32, 136
absorption and vaporization waves, synchronous propagation 32
absorption
   coefficient 10, 26, 30, 32, 35, 82–84, 121, 125, 135, 136
   halo 29, 31
   wave 28–30, 32, 34, 106
activation energy 36, 37, 39, 89, 91, 106
adiabaticity condition 55
Atwood number 58
avalanche breakdown 25, 29, 31

B

Bessel functions 132
Boltzmann equation 48
Bouguer law 103
boundary value problem 16, 26, 27, 37, 83, 89, 131

breakdown
   criterion 27, 31
   mechanism 25, 26, 29, 31
   optical 5, 28, 86, 106, 136
   of polymers 30
   probability 28
   threshold 28, 29, 31, 82, 136

C

chemical reactions 136, 137
condensation
   discontinuity 49, 50, 51
   shock 48, 49, 51
condition
   boundary 10–14, 19, 27, 32, 33, 58, 74–76, 83, 84, 89, 93, 110–112, 115
   hydrodynamic 45, 47–49, 116
   on ablation front 37
   on evaporatin front 16, 103, 116
   on discontinuity surfaces 51
   on melting front 15, 117
   on moving phase boundary 14
   on shock wave 51
   adiabaticity 48, 50
   supersaturation 47
conductivity, thermal 10, 27, 28, 36, 110
contact boundary 51, 57, 58, 135
convective instability 55
corrugation instability 6, 21, 73, 79, 88, 93, 101, 114, 128, 137
critical size of inclusion 28
current, photoemission 66

D

Debye temperature 10
deposition 63, 64
    of excited species 39
dispersion 28, 78, 80, 81, 94, 113, 118–121
dispersion equation 58, 76, 78–80, 86, 87, 89, 95, 112, 118, 120, 121, 129, 130
distribution function 28, 45, 46, 70
    in Knudsen layer 46
    of electrons 31, 69
    of inclusions 28
Drude-Zener, model 10, 19

E

electron
    relaxation dynamics 63
    temperature 64, 65, 66, 68, 69
    thermal conduction 28, 64, 65
electron-phonon coupling 64
emission
    anomalous 63, 70
    current 66, 67, 68
equations of gas dynamics 50
evaporation 5, 15, 25, 32–35, 45, 46, 48, 68, 73, 88, 97, 101, 103, 106, 109, 138
    front 16, 18, 21, 34, 74, 79, 80, 93, 115–118, 121
        velocity 16, 18, 74, 75, 115, 117, 119
        perturbed 76, 94
    rate 16, 18, 115, 117, 119, 120
    wave 16, 18, 32, 34, 80–83, 99, 100, 114
        in dielectrics 82, 101
excited species 36, 37, 39, 40, 88, 89, 91
expansion
    adiabatic 50, 55, 56
    of vapor 46, 47, 49–51
explosion 86, 87, 101

F

Fermi distribution 66, 69
Fermi energy 64
Fourier amplitudes 95, 97, 101
Fourier components 95, 96, 98
front
    of thermal wave 29
    vaporization 6, 19, 21, 22, 32, 45, 50
fusion 6, 15, 58, 136

G

gamma-function 53
gas-dynamics equations 49, 50, 52, 53, 55, 56
gravitational force 55, 57, 115, 119
gravitational-capillary waves 113, 114, 118
growth rate 76–78, 80, 81, 85–88, 95, 97, 100, 101, 109, 119–121, 130, 136, 137

H

heat conduction equation 10, 13–16, 19, 26, 29, 32, 36, 65, 74, 83, 93, 103, 110, 111, 115, 116, 126, 128, 132
hot electrons 65, 69
Hugoniot relations 48
hydrodynamic boundary conditions 45, 47–49, 116

I

inclusions 25, 26, 28, 29, 136
instability
    aperiodic 119, 120
    criterion 29, 91
    development 80, 113, 116, 137
    growth rate 56, 58, 78–80, 84, 85, 89, 91, 114, 117–119, 129
    hydrodynamic 9, 121
    of contact boundary 45, 58
    of laser vaporization 82, 93
    of vapor-plume expansion 54
    Rayleigh-Taylor 58, 59, 135, 136
    threshold 79, 100, 120

INDEX

interference 125, 127, 128, 129, 137, 138
inverse Bremsstrahlung 63

J

Jouguet condition 46, 49

K

kinetic equation 31, 45, 48, 69, 70, 127
Knudsen layer 45, 47, 48, 50, 51

L

Laplace
    operator 128
    transformation 11
        inverse 11, 12
latent heat
    of fusion 14, 15, 116
    of vaporization 15, 16, 47, 77, 83, 116
lattice, temperature 65, 68
Legendre polynomial 58
Lie group transformation 52
liquid 5, 9, 10, 14–17, 19–22, 31, 45, 69, 73, 79, 109, 110–116, 118–121, 137
luminescence 68, 69

M

Mach number 47, 51
Maxwell equations 19, 103
Maxwellian velocity distribution 45
melting and vaporization fronts 19, 20, 21
melting 5, 9, 25, 26, 35, 63, 114
    bulk and surface 15
    front 14, 19, 21, 116, 118, 120
        temperature 20
        velocity 14, 15
        kinetics at low superheating 14
    point 13–15, 20, 68, 73, 109
multiphoton ionization 9, 25, 29, 31

multiphoton photoelectric effect 65

N

Neumann's solution 14
Navier-Stokes equations 110, 111
nonlinear development of instability 73
nonstationary evaporation waves 79, 137
nucleation, homogeneous 15

O

optical absorption 9, 10, 25, 82
oscillating perturbations 88
oscillatory growth 79
oxidation 125–128, 132, 137
    kinetics 130
    law 128, 130, 131
oxide 30, 125–131, 136, 137
    layer 126–128, 130

P

periodic perturbations 94
perturbation, wavenumbers 94, 96
perturbations, short-wavelength 58
perturbed ablation front 89
perturbed evaporation front 76, 94
phase
    boundary 14, 17, 18, 20, 21, 34, 46, 50, 73–76, 79, 83, 84, 87, 94, 95, 97, 99, 100, 102, 104, 106, 111, 115, 116
    transition 9, 10, 135, 137
        in solids 15
        kinetics 14, 16
plume 54, 56, 57
    expansion 54
Prandtl number 112
principal radii of curvature 14, 74, 115
problem of Stefan 14, 16
pulsed UV-laser ablation 39

R

rarefaction wave 50, 51

centered 46
Rayleigh-Teylor instability (RTI) 58, 59, 135, 136
reflection 18, 104, 114, 127
reflectivity 5, 12, 18, 19, 20
relaxation time 37–39, 89–91

S

saturated vapor
    density 46, 47, 48
    pressure 13, 14, 16, 48, 50, 73, 75
Schrödinger equation 27, 84
seed electrons 31
shock wave 6, 50, 135
    compression 137
    generation 25, 56
    strong 45, 48
        structure 46
    velocity 51
similarity variable 49, 50
skin depth 10, 12, 13, 75
sonic point 46, 47
sound speed 14, 16, 46, 58
    in ambient gas 51
    in two-phase mixture 49
    in vapor 116
sound velocity 47, 119
spatial scale of temperature field 10, 12, 13, 17, 76
specific heat 10, 126, 131
    vaporization 5, 74, 75, 137
speed of light 19
speed of sound *see* sound speed
spin 100, 101, 138
spin sublimation 99
stability 17, 26, 74–77, 79, 83, 84, 88, 95, 97, 99, 100, 104, 114, 116, 128, 131, 132
    of ablation front 91
    of vapor expansion 54, 56
stabilization 58, 75, 87, 91, 120, 122
stationary
    evaporation wave 17, 18, 32, 33, 47, 73, 75–77, 80–83, 99, 106, 116

regime of evaporation 74, 79, 100
structures 99, 113
stimulated emission 36
stream function 114, 116, 117
strong
    evaporation 45
    shock wave 45, 46, 48
structure of vapor flow 45, 48
sublimation 79, 93, 99, 115, 118, 119
supercriticality 95, 97, 99, 100, 114, 137
superheating 14, 15
supersaturated
    state 47, 48
    vapor 50
surface
    energy density 4, 74
    evaporation 32, 115
    reflectivity 10, 12, 18, 20, 68, 125
    temperature 10–13, 18, 33, 35, 37, 39, 46, 47, 51, 64, 110, 115, 131

T

temperature field 10, 12–14, 16, 26, 29, 74, 79, 83, 93, 95, 99, 100, 104, 115, 128, 131, 132
thermal
    conductivity 64, 83, 103
    of dielectric 26
    explosion 125, 135
    instability 26, 135
    ionization mechanism 30, 31
    relaxation time 36, 88
    wave 26, 30
thermionic electron emission 5, 65, 66
thermocapillary
    instability 109, 110
    waves 112, 113
thermochemical instability 30, 136, 137

threshold 25, 28, 29, 69, 77–79, 93, 113, 131
　fluence for ablation 35
　intensity 27, 31, 82, 113
transparent dielectrics 6, 25, 31–33, 35, 82, 88
turbulence in vapor plumes 55

U

ultrashort laser pulses 6, 63, 65, 68, 70
UV-laser ablation of polymers 35, 39, 88

V

vapor 48–53, 59, 73, 83, 135
　expansion dynamics 47, 51
　plume 16, 45, 51, 54, 56, 58, 135
　saturated *see* saturated vapor
vaporization 9, 13–16, 34, 54, 69, 79, 87, 89, 96, 99–106, 114, 128
　front 73–75, 77, 83, 88, 95, 97–100, 104, 106, 115, 118, 137
　velocity 18, 32, 47, 104, 119
　wave 82, 83, 94, 95, 101, 103, 137
velocity distribution of evaporated atoms 45, 50

W

wave 17, 26, 29, 30, 32, 33, 70, 73, 75–79, 82, 100, 104, 114, 117, 119, 132, 138
　equation 19, 103
　length 77, 115, 120–122, 126–129, 131, 138
　number 78, 86, 87, 89, 93, 95, 97, 100, 101, 112
white noise 95, 96, 97, 98